Transformer Maintenance Guide

Third Edition

M. Horning
J. Kelly
S. Myers
R. Stebbins

Copyright © 2004 Transformer Maintenance Institute, Division of S. D. Myers, Inc. All rights reserved.

ISBN 0-939320-02-9

The Transformer Maintenance Guide is designed to provide accurate and authoritative information in regard to the subject matter covered. By using this maintenance guide the user agrees that the Transformer Maintenance Institute, Division of S. D. Myers, Inc., and its authors will not be held liable for any damages resulting directly from the use of information contained herein or from any errors or emissions.

This book, or any parts thereof, may not be reproduced in any form or by any electronic, mechanical, or other means, now known or hereafter invented, including photocopying and recording, without written permission of the publishers, S. D. Myers, Inc.

Printed and bound in the United States of America

Transformer Maintenance Institute

TMI (the Transformer Maintenance Institute) is the education and publishing arm of S. D. Myers, Inc.

In order to understand TMI, we must return to the origin of S. D. Myers, Inc. The company began as a knowledge-based operation. Before founding S. D. Myers, Inc., Stanley D. Myers worked at an electric supply company that also performed maintenance and service such as rewinding electric motors and small transformers. He noticed that, while the oil in new transformers was very light yellow in color and transparent, the oil in older transformers was sometimes so dark that you could not see through it. No one seemed to be able to answer his questions satisfactorily about this observed phenomenon. Therefore, he embarked on a quest for the answers. This quest led to the concept of extending transformer life by processing the oil. However, the company he worked for at the time was not interested in pursuing the idea. So, when Mr. Myers had the opportunity, he started S. D. Myers, Inc. with the objective of testing and processing oil in transformers to extend their reliability and overall lifespan.

Imagine starting a company to provide a service that no one knew they needed! The challenge became one of educating these potential customers. To accomplish this, Mr. Myers would share this knowledge with customers on sales calls and in sales literature. Once transformer owners began to understand the processes that worked to age and destroy their transformers, they knew that maintenance measures were needed to counteract the affects of this degradation.

Eventually, the knowledge that originally developed from continual study and research began to come from the experience of transformer operation, testing and maintenance. Over time the information that was being collected expanded beyond Mr. Myers' ability to share it through traditional sales literature and office visits with customers. To facilitate the communication of this knowledge to a broader audience, Mr. Myers founded the Transformer Maintenance Institute (TMI) in 1975. In that same year, TMI began producing the seminar, "An Introduction to the Half-Century Transformer, " which was (and still is) presented to customers in various industries all over the world. In 1981, only six years after TMI was born, they published the first edition of *A Guide to Transformer Maintenance*.

Since then, TMI has educated thousands, developed many other seminars and produced a myriad of technical papers. For information on our latest offerings, visit us at **www.sdmyers.com**.

Acknowledgements

This third addition would not have been possible without the help of many talented people.

First, we need to acknowledge all the people who were directed by our founder, Stanley D. Myers and put together (with primitive means) the first edition of the book in 1981. Looking back at all the effort that we put into the second and now third editions, we can't imagine how much work it must have been to write it from scratch.

In-house, we would like to acknowledge Nicki Lock for the excellent work on the text and graphic preparation.

A special thanks to Dana Myers, Al Cote, Dick Croghan, Chris Young, Matthew Gress, Ron Choat, John Sharnsky and Russ Williams, for their contributions and review of this material.

Special thanks to C. Clair Claiborne, Ph.D. Principal Consulting R&D Scientist, ABB Inc. Power Technologies Division for the Historical Background of Thermally Upgraded Insulating Paper and Thomas A. Prevost, Vice President Technical Service, EHV Weidmann Industries Inc. for the definition of thermally upgraded paper.

Thanks to Doble Engineering, Megger, and the Institute of Electrical and Electronic Engineers (IEEE), for pictures, figures and tables contained in this *Transformer Maintenance Guide*.

About the Authors

Mike Horning

After spending many years as a field service and electrical testing technician, in the office as a supervisor and manager of Field Services, Mike became a TMI instructor. Mike has been presenting seminars since 1996 and now is the Director of TMI. Main contributions to this book: Mike was the driver, organizing everyone's work to make sure we completed this project. He was the main conduit for updating the book's content based on his own experiences at the conferences, in committees, reading, as well as having the contacts throughout the industry to bounce ideas off of and gather information from. He was also responsible for writing a good many of the new sections of the second and third editions.

Joe Kelly

The Godfather of TMI. Joe presented the first seminar in 1976 in Houston, Texas with our founder, Stan Myers. Since then has been all over the globe for TMI. Joe has retired as Director of TMI. He still maintains a fairly full schedule presenting seminars and attending conferences. The list of articles Joe has authored or co-authored and the list of committees he is on or has been on would fill several pages. From the breadth and depth of Joe's experience he provided overall technical direction relating to new developments and insightful comments while reviewing our productions. His influence is still evident in this third edition.

Scott Myers

S. D. Myers Inc. was started in the basement of his family home. Scott grew up with transformer oil sample containers for toys. He worked summers at the company and started his first full time job right after college. Scott started in the lab, and then worked on a field crew for 3 years. He then quoted and scheduled fieldwork and as Vice-President managed all maintenance services. Scott wrote many of the new sections for the second edition that are included in this edition as well.

Randy Stebbins

Randy came to S. D. Myers, Inc. through the environmental side of our business having spent a number of years eliminating PCBs from electrical systems. He became Laboratory Manager in 1993 and started helping out as a TMI Instructor in 1996. He helped to write the current version of the TMI curricula and has taught seminars across North America and overseas. Currently Technical Director for S. D. Myers, Inc., he is responsible for all technical aspects of the company's activities as they relate to insulating liquids and chemistry. He contributed to the oil and chemistry chapters of the second edition and wrote several of the revised sections of the third.

Table of Contents

Transformer Maintenance Institute
Acknowledgements
About the Authors
Table of Contents...ix
Ask Charley...xix
History Of S. D. Myers, Inc..xxi

[1] Purchase..**3**
   Dryness..3
   Copper or Aluminum Conductors..4
   Disk vs. Layer Windings?..5
   Core Form Winding Construction..7
   Popular Types of Core Form Windings..7
      Layer Style..7
      Disc Style..9
   Load (Copper) Loss..13
      Controlling Load Losses..13
      Load Loss Conclusion..14
   Insulation Systems...14
      Dielectric Theory..14
      Winding Insulation..15
   Mechanical Force Restraints..16
   Thermal Characteristics..17
      Convection, Thermosiphon Flow, Radiation........................17
      Typical Types of Transformer Cooling..................................18
   Sizing...19
   Oil Preservation...19
   Upgraded Insulation...21
     Historical Background of Thermally Upgraded
     Insulating Paper...22
   Separate LTC Oil...24
   Basic Insulation Level (BIL)...24
   Standards..26
      ANSI/IEEE and IEC Equivalents for Transformer Oil.............30
   Factory Electrical Tests...36
   Site Installation..40
      Specifications..40
      Site Preparation..41
   Shipping...41
      Transformer Arrives by Truck...42
      Transformer Arrives by Rail..42

Internal Inspection..........43
Unloading and Transporting..........44
Field Assembly (Dress Out the Unit)..........44
Vacuum Leak Check the Transformer..........45
Vacuum Filling the Transformer..........46

[2] The Warranty Period..........49
One-Month DGA..........49
Ten-Month Full-Oil Test..........49

[3] Oil Testing..........51
Introduction..........51
Mineral Insulating Oil..........51
Ancient History of Transformer Oil..........52
The "Four Functions"..........53
What Causes the Oil to Age..........57
Accelerating the Rate of Aging..........59
    Oxygen Content..........60
    Heat..........61
    Other Accelerators - Moisture and Metals as Catalysts..........62
    Electrical Stress..........65
    Cellulose..........68
    Oxidation Products..........69
Controlling Accelerating Factors..........70
Developing a Testing Program..........71
    Evaluate Your Testing and Maintenance Program..........71
    When Do You Test?..........72
    What Do You Test?..........73
    What Tests Do You Run?..........74
Testing In-Service Transformer Oil..........78
    Liquid Power Factor (Dissipation Factor)..........78
    Moisture in Oil..........81
    Oxidation Inhibitor..........86
    Liquid Screen Tests - Oil Screen Package..........87
        Neutralization Number..........88
        Interfacial Tension Test..........91
        Relative Density (Specific Gravity)..........95
        Color..........96
        Visual Examination of Insulating Oil..........97
        Dielectric Breakdown Voltage D 877..........97
        Dielectric Breakdown Voltage D 1816..........99
    Monitoring and Diagnostic Tests for Transformer Oil..........101
        Dissolved Gas Analysis..........101

Dissolved Metals Analysis..102
Analysis for Furanic Compounds.................................103
Degree of Polymerization...107
Calculating DP and Remaining Life from
Furan Results..107
Analysis of PCBs in Insulating Oil................................109
Particle Count and Distribution....................................110
Particles and Filming Compounds Analysis...................111
Dissolved Gas In Oil Analysis by Gas Chromatography..........113
  Introduction..113
  History..113
  Prior to 1920..114
    The Buchholz Relay (1919)..116
    Field Gas Detectors...117
    Gas Chromatography Invented..117
    Sampling...118
  Test Description..121
    Gas Analysis by Chromatography..................................123
    Data Reproducibility..125
    The Physics of Gas in Oil..126
  GC vs. Other Methods..127
    Gas Collector Relay (Buchholz System)........................127
    Fault Gas Detection..129
    Gas Blanket Analysis..130
    Dissolved Gas Analysis..130
Test Data Interpretation...132
  Introduction..132
  Laboratory Demonstration..133
  Comparative Rates of Gas Evolution..............................135
  Qualitative and Quantitative Interpretation..................136
  Ratio Analysis Methods...140
  Case Histories Illustrating Basic Transformer Faults..........143
  Cautions in Analyzing GC Data......................................152
  Synopsis of the Procedure for Statistical Development
  of Surveillance Range Norms from a Gas Database.........153
  Comparing Current Methods of DGA Interpretation..........154
  Specifying and Testing New Oil.......................................157
    Avoiding the Pitfalls of New Oil Purchasing................157
    Suggested New Oil Purchasing Specifications............159
    Acceptance Testing of New Oil and New Installations......163
    Case History - But No Names, Please...........................165
  Tests to Run on New Oil from Suppliers......................166
  Testing Load Tap Changer Dielectric Liquid.....................167

New Testing Protocols..167
Traditional Oil Testing Methods.................................170

[4] Electrical Testing of Transformer Insulation.........177
Part 1 - Why Perform Field Electrical Tests?
    Introduction...177
    Oil Screen Testing..177
    Moisture-in-Oil (Karl Fischer)............................178
    Gas-In-Oil Analysis...178
    Thermographic Survey (Infrared)......................178
    A Complete Physical..178
Part 2 - A General Overview
    Records..181
    Safety..181
        General Precautions......................................181
        Entering Transformers..................................182
    Objectives..182
    Acceptance Tests (Factory & Field)..................183
    Periodic Tests...183
    Testing Units After Failure................................184
    Standards..185
    Application...185
Part 3 - Electrial Tests
    Guidelines..187
    Insulation Power Factor Series..........................188
    Winding Insulation Power Factor Test.............189
        Insulation Power Factor Tip-Up Test............194
        Bushing Power Factor Tests..........................195
        Liquid Insulation Power Factor Test.............197
        Single Phase Excitation Current Tests..........197
    Megohom Meter Series.....................................202
        Minimum Insulation Resistance....................204
        Dielectric Absorption....................................205
        Polarization Index...205
        Step Voltage Test...207
    Transformer Turns Ratio Test............................208
    DC Winding Resistance Test.............................210
    DC Overpotential Testing..................................212
    Core Ground...213
    Ground Resistance Measurements....................215
    Leakage Reactance Test.....................................216
    Frequency Response Analysis...........................217
    Recovery Voltage Method..................................219

[5] Transformer Coatings...221
Signposts of Paint Deterioration...221
    Solution to Paint Deterioration...222
    Preventive Versus Remedial Painting...223
    Surface Prepartion...224
    Scraping and Wire Brushing...224
    Powder Blasting...225
    Chemical Cleaning (De-Energized)...227
    The Primer Coat...228
    The Finish Coat...230
Methods of Applying Paint...232
Precautionary Measures...235

[6] Bushing Maintenance...237
Introduction...237
The Contamination Problem...237
Types of Contaminants...238
Dynamics of Contamination Collection...240
Theory of Contamination Flashover...242
Alternative Theoretical Apporoaches...243
The Flashover Process...243
Other Undesireable Effects of Insulator Contamination...246
Corona Manifestations...246
Corrosion of Metalwork...247
Visual Inspection...247
Preventative Maintenance...249
Hand Wiping...249
Periodic Insulator Washing...250
Dry Air-Blasting With Non-Abrasive Material...253
Use of Dielectric Compounds...255
Overinsulation...258
The Last Word on Cleaning...260
Internal Failure...261
Internal Failure Detection...261

[7] Transformer Life Extension...263
Introduction...263
What To Do If You Find Free Water In a Transformer...264
Transformer Leaks and Leak Repair...266
Electrical Tests to Determine Moisture Levels in
Solid Insulation...269
    The Power Factor Series...269
    The Megohm Meter Series...269

Dehydration of Transformers...271
   Introduction...271
   Dehydration to Avoid Failure...271
   Dehydration to Extend Transformer Life...272
   Insulation Drying Practices...272
   Factory and Repair Shops Methods...273
   Field Dehydration Methods...276
   Circulating Hot Oil...276
   Short-Circuited Windings, Vacuum...277
   High Vacuum...277
   Hot Air...278
   InsulDryer "On-Line" Transformer Dehydration Unit...280
Results of The Drying Process...280
   The Bonus of Degasification...280
   Drying Efficiency...281
   Dehydration and Degasification Only?...282
   Limitations of Dehydration Systems...282
Transformer Oil Regeneration and Hot Oil Cleaning...283
   Introduction...283
   Inadequacies of Existing Standards...284
   The Correct Oil Reclaiming (Regeneration) Strategy...285
      Oxidized Oil, No Visible Sludge Deposits...286
      Depleted Inhibitor, No Oil Oxidation Yet Detected...286
      Adequate Inhibitor Content...288
Starting From Nowhere...289
   How to Begin a Maintenance Program for a
System Suffering from Total Neglect...289
   Newer Transformers are Less Capable of
Surviving Neglect...291
   Which Transformers Would Cost the Most to
Replace?...291
   Delivery Time...292
   Significance to Your Operation...292
   Reviewing the Problem of Sludge...293
   Definitions...293
Confronting the Sludge Problem...295
   Introduction...295
   Seven Alternative Approaches...295
      Ignore the Situation...296
      Recondition the Oil...296
      Retrofilling the Oil...299
      Reclaim the Oil...299
      Hot Oil Clean® the Transformer...301

    Reinhibiting the Transformer Oil..................303
    Fluidex Technology..................305
Factors to Consider..................306
Energized Hot Oil Cleaning vs. De-energized
Hot Oil Cleaning..................307
When Will a Hot Oil Cleaned Transformer Require
Additional Treatment?..................308
Other Constraints of Energized Hot Oil Cleaning..................309
    Retrofilling or Inadequate Reclaiming..................309
    Oil Test Analysis..................309
    Observations..................311
    Transformer Age..................312
Recognizing that the Sludge Problem is Fixed..................312
The Reliability of Energized Hot Oil Cleaning..................312
Safety Precautions..................312
Energized Processing Innovations..................313
Energized Reclaiming - Objections Answered..................314
Visual Proof in Helping to Make a Valued Decision..................316
Conclusions..................317

[8] Remanufacturing..................**319**
Reusing Transformer Core Steel..................321
Failure Options for Large Shell Form Power Transformers..................322
Shell Form Power Transformer Corrected Problems..................324

[9] Disposal..................**327**
United Nations Environment Program - Persistent
Organic Pollutants..................328
Pure PCB Transformers..................331
Mineral Oil Transformers Containing PCB..................332
Mineral Oil Transformers Containing No PCB..................333

[10] Windings..................**335**
Purpose and Function..................335
Basic AC Theory..................336
Single Phase Transformer Polarity..................340
Three Phase Transformer Polarity..................341
    Phase Sequence and Angular Displacement..................341
    Phase Sequence and Angular Displacement of
    a WYE-WYE Connection..................342
    Phase Sequence and Angular Displacement of
    a WYE-DELTA Connection..................343
Types..................344

Design Criteria..344
All at a Minimum Cost..344
Materials..345
Conductor Shapes..346
Testing and Monitoring...347

[11] The Core...**349**
Purpose and Function...349
Brief History - Why Shell Form? Why Core Form?................349
The Magnetic Circuit..350
   Electromagnetic Induction..350
   Magnetostriction..351
   Core Noise..354
   Three Phase Core Form Core Construction....................355
   Core (No Load) Losses..357
   Conclustion (Core Loss)..358
History..359
Materials...359
Energy Efficient Transformers..360

[12] Solid Insulation..**367**
Stresses on Solid Insulation..367
   Dielectric Stresses..367
   Short Circuit Stresses...368
   60 Hertz Normal Excitation Voltage...............................372
   60 Hertz Transient Over Voltage....................................372
   Lightning Impulses..372
   Switching Surges..373
   Short Circuit Through-Faults...375
Solid Insulation Life..375
Conditions That Destroy Solid Insulation...........................377
   Water..378
      Factory and Installation...379
      Characteristics of Water Affecting Insulation..............383
      Oil and Water Do Mix!...385
      Chemically Bound Water...391
      Dissolved Water and Insulation Failure......................392
   Oxygen...399
   The Mechanics of Sludge Formation.............................400
   Heat..401
      Thermal Evaluation Testing.......................................403
      Unique Effects of Elevated Temperature on Aging.......404

Loading Guide for Oil Immersed Transformers............405
Effect of Loading Beyond Nameplate Rating................406
Transformer Insulation Life...407
The Rule of Eight Degrees...407
Nomex® Insulated Conductor.......................................408
    Thermal Considerations......................................409
    Electrical Considerations....................................409
    Mechanical Considerations................................409

[13] Cooling Systems...411
Purpose and Function..411
Types and History..411
Nameplate Designations..414
Adding Additional Cooling Capacity.............................415
    Due to High Ambient..415
    To Increase Transformer Capacity.....................415

[14] Tank and Paint...419
Purpose and Function..419
The Corrosion Problem..420
Dampness - The Overriding Factor in Corrosion...................421
Selecting the Right Coating System...............................422
    Environmental Criteria.......................................422
    Effect of Color on Paint Systems.......................422
    Exploring High-Performance Coatings..............426
    Ongoing Research...427

[15] Bushings..429
Purpose...429
Design...429
Materials...430
What Could Go Wrong...430

Appendix A - Acronym Indentification.........................433
Index...435

If, after reading this book, you have a question on a particular subject, just **Ask Charley.**

A feature of the S.D. Myers website, Ask Charley allows visitors to our site to take advantage of the vast amount of technical information we have assembled concerning the maintenance and operation of substation equipment. This is a place where customers can ask general questions of our team of experts and receive general answers relating to substation maintenance and operation, view the questions and answers asked by other peers in your industry, and network with your peers via the Customer Forum (in a message board format).

We hope you will enjoy this feature. Just point your browser to: www.sdmyers.com and look for the Ask Charley button.

History of S. D. Myers, Inc.

Shortly after World War II, Stanley D. Myers joined a small electric supply company and motor rewind shop. As this company grew, they were contracted to service and rewind many sizes and types of transformers. Noticing the varying conditions of the oil in these transformers, it became apparent to Stan that if the oil could be kept in better condition, the result would be a more reliable transformer. His interest in this area led him to research and investigate any material that had already been printed on the subject and transformer evaluations by major manufacturers, electrical rewind shops, and professional organizations.

His research then led him to design and develop a piece of equipment called a "Re-Refiner" which would remove moisture, gas, acids and sludge from the oil. In 1965, Stan received his first purchase order for this process and "The Transformer Consultants" was born, later to become S. D. Myers, Inc. (SDMI). A short time later, Mr. Myers developed a process for "re-refining" that could be done safely while the transformers remained energized.

By 1967, one problem had become apparent: poor oil test data. The causes were found to be the owners' improper identification of their transformers and poor sampling technique. To solve this problem, Stan hired and trained the first sampling and testing technician and supplied him with a mobile testing laboratory. This individual would sample the units, test them on site, and discuss these results with the owner before leaving. If the oil needed service, SDMI generated a proposal to solve the customer's problems.

By 1972, the company began to grow rapidly by simply listening to customers' concerns and needs and responding accordingly. That same year, SDMI expanded their offerings to include energized insulator cleaning and painting of transformers.

By 1974, just nine years after the company began, the fleet of oil testing and servicing operations had grown to five oil "Re-Refiners," now called Reclaimers, and eight Mobile Test Laboratories (MTL's) serving industries and utilities nationwide.

1975 saw the birth of the Transformer Maintenance Institute (TMI) and captive field Substation Electrical Testing capabilities. TMI is the company's education and training arm, and provides valuable information on substation and transformer maintenance through seminars, training, publications, and consulting around the world.

In order to stay current and share company experiences, Mr. Myers and an associate, Joe Kelly, became active members of the Institute of Electrical and Electronics Engineers (IEEE), the American Society for Testing Material (ASTM), and other key technical organizations. Because of the rapid increase in knowledge concerning transformer oil evaluation and the improvements in the laboratory equipment, the company created its first in-house central laboratory in 1977. Because dissolved gas analysis and moisture-in-oil tests required delicate equipment not suited for mobile operations, on-site testing for the oil screen tests was nearing an end. Today, the company has a team of highly trained technicians that continue to provide substation inspections and obtain dielectric fluid samples for testing from tens of thousands of units per year. Testing of both new and in-service insulating liquid and paper samples continues to grow more sophisticated and provide more useful and practical maintenance related information.

In 1978, the company began to manufacture its first engineered product, "Cool-a-Tran," which is an after-market device that is designed to efficiently dissipate unwanted heat from transformers. Cool-a-Trans continue to be produced today for those special applications.

The 1980's were an exciting time for SDMI and provided many challenges and opportunities. TMI published its first book, *A Guide to Transformer Maintenance*. The 835-page book was a concentrated effort to share knowledge exclusively on fluid-filled electrical equipment. The book became a worldwide standard for many transformer owners, insurance companies and universities. In 2001, the book was updated and a second edition published. And, now, in 2004, TMI introduces the third edition of this updated classic.

Also during the 1980's, SDMI's Manufacturing division expanded and built specialized oil reclaimers, as well as small Mobile Cartridge Filters (MCF's).

The most significant challenge to the company in the 1980's was complying with TSCA – the Toxic Substances and Control Act – that had been passed in 1976 and became effective in 1979. TSCA includes a ban on manufacture and import, as well as associated guidelines regarding the disposal, control and use, of Polychlorinated Biphenyls (PCB's). SDMI's laboratory worked with others in the industry to develop a standard for testing mineral oils for PCB's. In the latter part of 1980, SDMI started retrofilling oil filled transformers to reduce PCB levels. From the beginning, it was apparent this process was very limited. For large-gallon units and units with high-concentrations of PCB's, retrofilling was not an economical alternative. The company developed and sought EPA approval for a process that would chemically destroy the PCB's in oil. In 1982, SDMI received EPA approval to use this process at both customer sites and at our central facility. In keeping with our philosophy of processing transformers while they are energized, "PCB-GONE" not only safely destroyed the PCB's but also reclaimed the oil to a like-new condition specification.

By 1985, SDMI had grown out of its original facility and moved to a much larger one in Tallmadge, Ohio, where we still maintain our corporate headquarters. Just 2 years later, our transformer repair and rewind department was born.

The PCB laws concerning the approved methods for disposal of PCB-filled equipment were very limited. Incineration of solids, liquids, capacitors and burial of the carcasses did not provide owners of PCB transformers many options. Some became victims of unscrupulous vendors, were subject to fines and financially responsible for the clean up and disposal of their PCB's a second time.

Identifying this problem, SDMI developed an alternative, more secure, method for the disposal of PCB transformers. In 1989, the company obtained the first EPA approval for recycling transformers. The solid insulating materials and dielectric liquids were incinerated, but the case, core, and conductor metals were decontaminated and recovered for recycling. By 1992, approvals had been granted to apply this technology to capacitors, lighting ballasts, lead cable, and several other types of PCB and PCB-contaminated equipment.

Consistent with continual development of improved laboratory capabilities, SDMI personnel played a key role in standardizing furanic compounds analysis. Starting in the early 1990's and continuing today, SDMI has contributed to both the ASTM and IEC standards for analysis and have been pioneers in using and interpreting furan results.

In the 1990's, SDMI expanded beyond the borders of the United States. Partnerships and operations were established in Australia, Saudi Arabia, Mexico, Canada, South Africa, South America and Morocco.

The turn of the century has been a time of acquisitions. In 2001, SDMI purchased Sun-Ohio, a competitor in the area of chemical destruction of PCB's in oil. While this market has matured domestically, S. D. Myers International continues to provide this service in other parts of the world.

The latest acquisition, completed in March of 2002, was the purchase of Ohio Transformer, a major player in the business of rewinding and remanufacturing transformers. With Ohio Transformer, SDMI has the experience, personnel and equipment to repair or rewind/refurbish small transformers and regulators at their Tallmadge, Ohio facility, medium-sized units at their Louisville, Ohio facility and large transformers at their Bradenton, Florida facility.

Lastly and most importantly, S.D. Myers, Inc. was founded on Biblical values. Stan Myers was a man who believed that the only way to attain salvation was through Jesus Christ, the Son of God. He also believed that as Christians it is our responsibility to tell others about the salvation God so freely offers. This remains *the* guiding principle of our company.

"There is no salvation in anyone else, for there is no other name in the whole world given to men by which we are to be saved."

- Acts 4:12

Chapter 1 | Purchase

Actions to take for assuring maximum, reliable operating life for your transformers begin with the specifications for ordering a new transformer. The electrical engineers designing your system will define a number of requirements for transformers to serve and function properly in that system. In addition, if you pay attention to the following considerations you can set the stage from the beginning for the long, reliable life for your new transformers.

Dryness

The moisture content of a transformer's solid insulation plays a major role in determining a transformer's length of life. Every time the moisture content of the solid insulation doubles, the expected life of the transformer is cut in half. Throughout a transformer's operating life, moisture will accumulate in the solid insulation. This moisture originates either from: outside the transformer or from within, as the liquid and solid insulation age and oxidize over time.

Each transformer begins its operational life with a certain amount of moisture in the solid insulation due to the manufacturing process - both of the insulation itself and of the transformer. In humid shop conditions, solid insulation exposed to atmosphere can absorb as much as 8 or 9% moisture by dry weight (%M/DW). Since solid insulation will absorb more moisture from the atmosphere than anyone wants to have in an operating transformer, manufacturing and repair shops dry the core and coil assembly using heat and/or vacuum, before tanking and filling the finished transformer.

The prospective buyer needs to specify the level of dryness in a new transformer so as not to just leave this very important quality to the whim or production schedule of the manufacturer.

A specification value of 0.5% maximum moisture by dry weight is recommended for new equipment. Manufacturers can achieve this level, and frequently dry to even lower levels than the specification limit. At times it may require a little extra drying time in the oven, possibly interfering with the manufacturer's production schedule, but as the buyer you have every right to specify the level of dryness you require for your new transformer. Keep in mind that just an extra half percent, for a total of 1.0 % M/DW, will cut your transformer's expected operating life in half. A little extra time in the oven is certainly worth the extra effort!

Of course, after you take delivery of the transformer, the responsibility becomes yours to make sure that your transformer's insulation stays dry. If you don't, then the extra work you specified from the manufacturer becomes wasted effort. For more on moisture in solid insulation see "water," beginning on page 378.

Copper or Aluminum Conductors?

A transformer is a piece of production equipment. Therefore purchasing, maintenance and repair decisions must be made in order to spend the least amount of money while still yielding the most performance out of that piece of equipment. However, nearly everyone knows that the lowest cost option up front often is the higher cost option over time. Such is the case with aluminum windings.

Aluminum has the advantage over copper of being less expensive to purchase per pound. That's pretty much where the advantages end.

Copper is a much better conductor than aluminum (see Table 10.1), so in order to carry an equal amount of current, an aluminum conductor must have a greater cross section. This translates into a larger coil and often an overall larger transformer. (Which coil weighs more? Which has greater resistance?)

Aluminum also has a lower melting point (660 °C) and less mechanical strength than copper. This means that a copper wound transformer can withstand much greater stresses before possible failure than can a comparable aluminum wound transformer.

Adding it all up, the small additional cost of a copper wound transformer more than pays for itself over the life of the transformer.

Disc or Layer Windings?

In dealing with core form construction, layer-wound transformers present another option to save money during the manufacturing process. But, the money saved purchasing layer windings rather than disk windings will most likely be lost from reduced transformer life.

This is beacause in a layer-wound coil, each turn consists of a sheet of conductor which is as wide as the coil is long or of conductor material which is wound continuously the length of the coil. Disc coils consist of individual strands of conductors and provide much greater mechanical strength than layer wound coils.

Figure 1.1 - Layer winding (cross section)

Figure 1.2 - Disc winding

Figure 1.3 - Failed layer winding - A disc wound transformer probably could have withstood the forces that caused this layer wound transformer to fail. Notice how the core bracing couldn't hold the layers in place.

Core Form Winding Construction

A core form winding is a cylindrically shaped coil, consisting of insulated conductive turns and various other insulation materials placed between the turns and ground. The coil is mounted over a leg of the magnetic core with its vertical axis.

The goal is to provide adequate effective dielectric strength to withstand operating voltage, impulse surges, switching surges, fault currents and test voltages. Adequate coil ventilation to meet thermal requirements and adequate mechanical strength is also required. This all needs to be done at a minimum cost.

Popular Types of Core Form Windings

Layer Style - cylindrical coil construction of one or more layers of insulated turns with pressboard layer insulation and vertical cooling ducts.

Low Voltage Layer Application - used for voltages up to approximately 15 kV and has no serious current limitations. These can be wound with an even or odd number of layers to accommodate electrical circuit connections of the start and finish coil leads. These coils are frequently wrapped with thermo-setting polyester glass tape which hardens during the baking process to mechanically bind the coil.

Regulating Voltage Winding Layer Application - transformers equipped with load tap changers (LTC's) generally use a separate regulating winding to provide the tap voltages. This layer winding is constructed so that the turns between each LTC tap are distributed over the entire length of the winding. This technique "electrically balances the winding" to better withstand short circuit forces.

High Voltage Layer Application - General Electric developed layer windings for use in large capacity high voltage generator step-up and autotransformers. This application was used up to 1200 MVA and to 2050 kV BIL (765 kV) single and three phase. ABB also used high voltage layer windings. Due to the geometry of the turn, these windings have excellent transient (impulse) voltage distribution with their inherent high series capacitance. They also have lower reactance, which minimizes the effects due to the voltage regulation drop through the transformer. This allows for maximum power transfer without affecting the stability of the power system.

The majority of large core form transformers use continuous transposed cable (CTC) for the winding conductor. The individual conductor cross section is small to limit the eddy currents and stray losses. Each conductor in the bundle is insulated with enamel film and an epoxy bond. The position of the conductor is continuously shifted along the length of the cable bundle to control circulating currents caused by stray flux fields. The cable bundle is wrapped with high strength Dennison crepe paper insulation that forms the turn insulation.

Winding cylinders for high voltage layer style coils are composed of numerous layers of Kraft paper bonded with resin. These cylinders are placed between the inner winding and the core and between winding layers for mechanical support and electrical insulation. Coil supports in the way of end rings are used at the ends of layer windings. The rings are cut from cylinders and one end is tapered to match the pitch of the winding.

The remainder of insulation between layers for high voltage layer style windings is formed by hard pressboard strips attached to a backing sheet of Kraft paper. The backing sheet and pressboard strips are wrapped on the coil turns between layers to form oil ducts which function both as electrical insulation and as cooling for the winding conductors.

A cylindrical electrostatic shield is placed immediately adjacent to the HV line lead and the line end turn of the winding to limit transient surge voltages on the first several turns of the winding.

A shielded turn (non-active turn) can also be added between turns at the line end or placed between turns around the tap coils if additional surge protection is required.

This style of high voltage layer winding is wound on a horizontal lathe with special axial compaction equipment. The construction quality of these windings is very unforgiving and great care must be exercised to assure the accurate placement of the winding turns.

Disc Style - cylindrical coil constructed as a series of pancake coils with insulating radial spacers separating the pancakes

Spiral (Helical) - this construction is used for voltages up to about 15 kV and approximately 3000 amperes. The spiral winding is a disc winding with "one turn per disc." The helical winding has a pitch where the continuous disc winding does not. Surge distribution is not a serious design concern for this type of winding since they are used in low voltage applications.

The turn normally consists of two groups (stacks) of conductors wound side-by-side in parallel. The turn is wound over axial spacers that provide a duct between the coil and winding cylinder.

Keyed radial spacers, locked in place by vertical spacers, are used between the turns (discs). The turn conductors are transposed as the winding progresses so that each conductor is shifted in position through the turn bundle at least one time from the start to the finish end of the coil. This procedure limits stray and eddy losses in the winding.

Continuously Wound Disc - This style of disc winding is commonly used for higher voltage windings up to about 450 kV BIL and approximately 300 amperes. These windings are fairly simple, compact, uniform and require minimum electrical clearances. They have a non-uniform but predictable surge voltage distribution. The winding turn generally consists of one to four copper straps in parallel. One disc constitutes a number of specified turns.

The concept of the continuous wound disc is to completely wind the coil from start to finish without a brazed joint. The initial turn will be wound around the insulating cylinder, and then seven turns will be wound over the initial turn. The stack of turns is then repositioned (flopped) so the "start" of the winding (line lead) will be at the outside (turn 1) and turn 8 will be located at the inside of the first disc. Turn 8 is then bent and formed as an inside crossover connection and becomes turn 9 at the beginning of the second disc. This disc is wound straight up to turn 16. Turn 16 is then bent and formed as an outside crossover connection and becomes turn 17 at the outside of the third disc. The third disc of 8 turns is wound on a let down ramp fixture and these turns are repositioned (flopped) so that turn 24 is located at the inside. This winding procedure is repeated until the coil is completely wound. Starting with disc 1, each odd numbered disc requires repositioning of the turns. Each even numbered disc is wound straight up. The last disc of the winding will have the last turn (line lead) at the outside.

High-density radial spacers keyed to axial spacers provide the duct space, which separate the discs for cooling and insulation purposes.

High voltage disc winding application - for disc windings above 450 kV BIL, "non-uniform" surge voltage distribution introduces design problems for engineers.

By definition, capacitance is the property of a system of conductors and dielectrics, which permits the storage of electricity when potential differences exist between the conductors and the conductors to ground. We can equate the two plates, the distance between them and the distance to the inside of the box as our disc windings, the insulation between discs and insulation of the discs to ground. C_s is called the series capacitance (capacitance between winding turns and discs) and C_g is called the ground capacitance (capacitance of the discs to ground).

If we had an ideal disc winding, the surge voltage distribution across the winding would be a straight line connecting 100% of applied voltage at 0% of winding turns (line end) to 0% of applied voltage at 100% of winding turns. This is called linear distribution and does not occur in disc windings because the actual voltage distribution stresses the line end of the winding. This condition exists because of the capacitive effect of C_g/C_s and is referred to as the alpha (α) of the coil.

We can see that a typical continuous winding (450 kV BIL and lower) has an α = 10, and 65% (100% - 35%) of the applied surge voltage is distributed across only 10% of the line end turns of the winding.

To improve (lower) the coil α = C_g/C_s, we can increase C_s (series capacitance of the winding) by interleaving the line end turns of the discs. This style of disc winding used for high voltage winding application (above 450 kV BIL) is called a Hisercap (high series capacitance).

Using Hisercap construction lowers our coil to 1 and 65% of the applied voltage is now distributed over 60% of the total winding turns. It is less risky and more economical to design disc windings having lower coil alphas approaching 0, which is the ideal condition for the higher voltage disc windings.

Another perhaps more popular method of increasing the series capacitance C_s and lowering the alpha of a disc winding is by using a shielded turn construction in the line end discs.

The electrically floating shields are inserted within pairs of discs with a formed crossover between discs. The number of shields per pairs of discs is determined by the winding BIL. As the BIL is increased, more shields are used. The shield ends are terminated with special dead end taping after completing a full or fraction of a turn.

This design results in excellent surge voltage distribution along the coil length and provides a low alpha.

The shielded turn does not require conductor brazing as needed at the outside crossover for the Hisercap construction. This simplicity of construction assures high quality and service reliability.

As discussed in the High Voltage Layer Application section, shielded turns are used as surge suppression devices in a similar fashion as disc windings. Shielding is used in the first turns of the line end of the coil and around tap turns to control voltage stress.

Load (Copper) Loss

During the operation loading of a transformer, currents flow in the high voltage and low voltage windings. An electromagnetic field results during this process, which produces load flux or leakage flux. This phenomenon causes load (copper) loss.

The transformer load loss is comprised of three components:

- I^2R loss, which is due to the current flowing through the resistance in the windings.
- Eddy and circulating current losses are caused by induced currents in the winding conductors resulting from the electromagnetic field and uneven length of the winding turn.
- Stray losses, which are a result of the load flux penetrating the clamping frames, tank wall, core, shielding etc. attempting to complete a path.

Controlling Load Losses

Adding more cross sectional area to the turn conductors will reduce the resistance and therefore the I^2R.

Subdividing the turn conductors and transpose them will reduce the eddy and circulating currents. The eddy currents are directly proportional to the square of the conductor width and transposing the conductors repositions them in the electromagnetic field.

Stray losses will be reduced by moving structural parts further away from coils, providing adequate magnetic shielding and using non-magnetic metallic parts will reduce stray losses.

Load Loss Conclusion

Core loss (p. 357) exists immediately and is a constant value when the transformer is energized.

Load loss (as the name implies) begins when the transformer is loaded and increases with the square of the load that the transformer sees. Thus, the load loss at 25 MVA of a 50 MVA rated transformer is $(25/50)^2$ or only 1/4 of the 50 MVA load loss.

As with core loss, load loss is an expense to the power company. Since the load loss varies widely with the transformer load, an average evaluation for this expense is $1200 to $1500 per kilowatt of loss.

Insulation Systems

Insulation and mineral oil are perhaps the most important parts of a transformer. These items are mainly responsible for preventing the transformer from destroying itself and other associated equipment.

Dielectric Theory

Transformer insulation systems consist of various forms of pressboard materials in combination with mineral oil. Both pressboard material and oil have limits on their dielectric withstand capabilities. Oil is the weaker of the two materials and generally limits the insulation design.

An optimum insulation system consists of small pressboard barriers with larger oil ducts to limit the stresses within the insulation system to safe levels. A designer's goal is to design an efficient, economical insulation system that limits or prevents partial discharge. Partial discharge is unwanted electrical discharge within a transformer winding or insulation system that occurs when the electric-field intensity exceeds the dielectric strength of the insulation structure.

Winding Insulation

Minor Insulation (within windings)
Conductor insulation is generally one or a combination of Dennison Crepe tape, Kraft papers and Nomex. These are used between winding turns, as layers and disc turn filler, as radial and axial spacers and pressboard wraps. Thickness and placement increase with voltage level (BIL).

Major Insulation (external to windings)
The major insulation is considered to be the insulation between windings, pressboard barriers, oil ducts, formed pressboard angles and winding caps.

Major insulation is also between windings and ground, the pressboard end rings, lebonite blocking, winding cylinders, pressure rings, phase-to-phase insulation, and formed pressboard parts.

Insulation Design Analysis (Minor and Major Insulation)
Transient (Impulse) Voltage computer program calculates the impulse voltage distribution through the winding. This enables the designer to make the most efficient use of the winding (minor) insulation and results in smaller coils and reduced copper weights. The designer is able to add copper to lower losses or increase winding BIL level during remanufacturing.

Electric field plot is a computer program used to determine electric field gradients at various points within the winding insulation structure.

The design engineer can compare electric field gradient with design limits of the insulation and modify insulation design to prevent voltage stress concentrations. This promotes the most efficient and economical use of the insulation system.

Mechanical Force Restraints

The internal structural members of the transformer must be designed to control stresses of mechanical forces. These structural members include:

- Top and bottom core side frames
- End frames holding the top and bottom side frames together
- Lock irons and bearing plates
- Top and bottom gusseted blocking skirts
- Jackscrews/pads and a 360-degree clamping ring
- Top and bottom coil end blocking
- Internal bracing to the tank wall

Mechanical forces involved:

Static Loads - weight-related and without motion
- Clamping the core to keep it from falling apart
- Supporting the weight of the coils
- Mounting of the terminal board (superstructures)

Dynamic Loads - produce physical forces in motion
- Forces occurring during transformer overloading
- Substantial forces when an external fault to the transformer occurs
- Forces that occur during transportation and seismic events

Many power transformers built or remanufactured today use two styles of mechanical force retraining schemes:

Static Restraint uses a 360-degree clamping ring (pressure plate) with jack bolts and discs. These vertical forces are transmitted uniformly into the coil blocking under the coils and over the coils and also to the 360 degree clamping ring at the top of the coils.

Dynamic restraint uses spring loaded clamps called "DYNA-COMP." This device consists of a large spring in an oil filled cylinder with a piston that exerts pressure to the top coil support plate. This assures that pressure is applied for static loads and a spring-loaded dashpot effect for dynamic loads such as axial short circuit forces or the shocks of transportation.

The DYNACOMP device is designed not to lose coil-clamping restraint over the service life of the transformer even if the coil insulation members shrink due to aging or additional moisture removal. The Dynacomp system is more expensive than the static clamping ring style. It is used only on large core form power transformers, usually autotransformers.

Thermal Characteristics

Energy losses in a transformer appear as heat. This heat must be dissipated without allowing the windings to reach a temperature that will cause excessive deterioration of the insulation. The transformer cooling is a means of transferring heat from the oil to the surrounding atmosphere. The cooling is designed so the oil and average winding temperature does not exceed a 65 °C rise, and the winding hot spot temperature will not exceed an 80 °C rise.

Convection, Thermosiphon Flow, Radiation

The heated oil becomes lighter and rises, creating a column of hot oil. Cooler, heavier oil next to the tank wall and bottom replaces the rising heated column of hot oil during the process of convection. A circulation of oil from the top to the bottom begins and is called the thermosiphon flow of oil. Dissipation of heat by radiation takes place because the tank and radiators are raised to a temperature above their surrounding environment.

Typical Types of Transformer Cooling

Cooling schemes vary depending on requirements. Some cooling styles use oil pumps, cooling fans and barriers to direct the oil flow.

OA (ONAN) = Oil Immersed Self Cooled
Radiators are added to the tank to increase the tank area thereby dissipating the heat by radiation. The temperature difference between the top of the tank and radiators (hotter) and the bottom (cooler) promotes convection and the thermosiphon flow of oil.

OA/FA/FA (ONAN/ONAF/ONAF) = Oil Immersed Forced Air Cooled
This is the same as OA cooling except cooling fans are added to the radiators. Supplemental kVA ratings of 133% and 167% of the OA rating can be obtained by placing one half or all of the fans in operation. The average oil rise at the supplemental kVA ratings is reduced allowing the average winding rise to remain below the 65 °C guarantee

OA/FOA/FOA (ONAN/ODAF/ODAF) = Oil Immersed Forced Air / Forced Oil Cooled
Same as OA/FA/FA cooling except oil pumps are used to force the oil through the transformer and radiators at an increased rate while the fans are cooling the oil.

Barriers inside and around the coils may be used to guide and direct the oil flow into the windings.

The addition of the oil pumps and barriers (directed oil flow) can achieve the same supplemental kVA ratings as OA/FA/FA with less radiators and cooling fans.

Sizing

When specifying a transformer to purchase, the owner should be mindful of the load that transformer might be called on to carry - not just upon installation but also over its operating life. If the unit will likely be called on within a few years to carry a load 25% or 50% greater than its load at installation, make sure to size it for the larger, future load. Why is this a concern?

The load a transformer carries translates directly into its operating temperature. Its operating temperature, in turn, translates directly into its reliable operating life. Heat, among many other factors, directly determines the life of a transformer's solid insulation. For every 8 °C that the winding's operating hot spot temperature increases, its reliable operating life is cut in half. For every 8 °C decrease, the insulation's life is doubled. (See The Rule of Eight Degrees, p. 407) Therefore, making sure that new transformers are sized to carry potential load increases will significantly increase their reliable operating lives! Don't neglect this consideration when specifying a new transformer.

Oil Preservation

Transformer oil serves four functions in an operating transformer. The most important of the four functions is to protect the solid insulation. (See The "Four Functions," p.53) If the oil provides a ready channel for oxygen and moisture in the atmosphere, giving it ready access to the solid insulation, then it has failed at this very important function. Additionally, ready access to oxygen and moisture allows the oil itself to deteriorate at an unacceptable rate. Deteriorating oil also prematurely ages solid insulation. (See Transformer Life Extension, p.285)

Transformer manufacturers offer the buyer several options to preserve the oil from contact or interaction with atmospheric oxygen and moisture. Be familiar with all of the options your manufacturers are offering and their prices. Choose the oil preservation system(s) best suited to your needs and budget.

An active, nitrogen system is by far the most effective oil preservation system presently available. Such a system includes a nitrogen bottle, a regulator in-line between the bottle and the transformer, and a pressure relief valve on the transformer. If the transformer is properly vacuum filled at installation, an active, nitrogen system nearly eliminates the effects of oxygen on both solid and liquid insulation. If the size of your transformer, the price of the system and your budget allow it, this is the best way to go.

A sealed oil conservation system is another very effective (and still expensive) system. This system includes a conservator tank with a flexible bladder that allows oil to expand and contract with changing operating temperatures while physically sealing the oil from the atmosphere. As with the previous nitrogen system, the cost of this system makes it a viable option mainly for larger transformers.

Substation class transformers are often physically sealed against the atmosphere with a gas blanket (usually dry nitrogen) above the top oil level. The gas blanket allows for expansion and contraction of the oil over the range of operating temperatures. When properly installed, a gas blanket system can remain intact and effective for a number of years. The integrity of this system can be breached by the smallest, pinhole leak, though. Oil in these systems should come with 0.3% oxidation inhibitor to guard against such failures. (see the discussion of Oxidation Inhibitor, p.288)

Some substation class transformers are designed to be free breathers, that is, they are open to atmosphere to accommodate the expansion and contraction of oil during operation. A transformer of this configuration should always include a desiccant breather to minimize the amount of moisture entering the transformer. Since free breathing transformers don't prevent oxygen from the atmosphere from entering the transformer a different strategy must be employed to counter the harmful effects of oxygen. These units require 0.4% oxidation inhibitor in the oil. The oxidation inhibitor in these smaller, less expensive units provides some of the protection against oxidation that more expensive oil preservation systems do for much larger transformers.

Upgraded Insulation

In the 1960's, Kraft insulating paper upgraded by dicyandiamide began appearing on the market in North America. The upgrade with dicyandiamide allowed transformers to be operated with a 65 °C heat rise versus the previous 55 °C heat rise achievable with non-upgraded insulation. Dicyandiamide not only allows hotter operation of a transformer, it also makes the solid insulation much more resistant to damage by heat, oxygen, water and oil oxidation by-products. This doesn't mean that owners of transformers with upgraded insulation can neglect them, allowing heat, oxygen, water and oxidation by-products to build up in their transformers. Rather it means that proper maintenance practices provide even greater benefit in terms of added years to a transformer's reliable, operating life.

Practically all new transformers in North America since the 1970's are 65 °C rise units due to the addition of dicyandiamide to the solid insulation. Outside North America many if not most transformers still are rated at 55 °C rise without dicyandiamide upgraded insulation. If upgraded insulation is available be sure to specify it. If your manufacturer doesn't offer upgraded insulation yet, ask them to add it to their product offerings.

*Thermally upgraded paper is cellulose based paper which has been chemically modified to reduce the rate at which the paper decomposes. Aging effects are reduced either by partial elimination of water forming agents (as in cyanoethylation) or by inhibiting the formation of water through the use of stabilizing agents (as in amine addition, dicyandiamide). A paper is considered as thermally upgraded if it meets the life criteria as defined in ANSI/IEEE C57.100; 50% retention in tensile strength after 65,000 hours in a sealed tube at 110 C or any other time/temperature combination given by the equation:

$$\text{Time (hrs)} = e^{(15,000/(T+273)-28.082)}$$

Because the thermal upgrading chemicals used today contain nitrogen, which is not present in Kraft pulp, the degree of chemical modification is determined by testing for the amount of nitrogen present in the treated paper. Typical values for nitrogen content of thermally upgraded papers are between 1 and 4 percent when measured in accordance with ASTM D 982.

*Note: This definition was approved by the IEEE Transformers Committee Task Force for the Definition of Thermally Upgraded Paper on October 7, 2003.

Historical Background of Thermally Upgraded Insulating Paper

The first transformers ever to have stabilized insulation made use of an amine compound. This method employed a reservoir or in some cases a bag containing an amine compound which was slightly soluble in oil. Theoretically, this amine compound (dicyandiamide) migrated to the paper and stabilized hydroxyl radicals preventing them from associating with hydrogen ions to form water and initiate a breakup of the cellulose chain by hydrolysis[1]. In actual practice, this method did not perform satisfactorily, probably due to the inability to get complete saturation of the insulating paper. In fact, it was fairly quickly concluded that little if any improvement of the paper was achieved by this method.

Subsequent improved forms of preparing the paper during its manufacture were much more successful. One type of thermally upgraded paper is made by treating regular insulating paper with a combination of dicyandiamide, melamine and polyacrylamide.

This is the Westinghouse (and subsequently ABB) method known as Insuldur.[1]

GE and McGraw apparently each had their own methods The GE method called Permalex used cyanoethylated acrylonitrile and the McGraw-Edison method was called Thermecel, using protein, p-aminophenol, and morpholine. It is difficult to find anyone who just used cyanoguanadine.

These methods all relied on the measurement of nitrogen in the finished product to determine the degree of upgrading. Now this gets somewhat complicated. The method referred to is ASTM D 982. This method was last reapproved in 1971 and went obsolete in 1980 and was not replaced. However the method is relatively current in TAPPI, the Technical Association of the Pulp and Paper Industry, Inc as T418 (1997).

Apparently, the paper manufacturers still use one of these methods. The method is a wet chemistry method called the Kjeldahl procedure. You digest the compound (paper) in concentrated sulfuric acid, add sodium sulfate and mercuric oxide and then measure the nitrogen as ammonium sulfate in the excess acid.

According to specification, the nitrogen content should be around 1.3 to 2.6 % nitrogen (according to the Westinghouse/ABB specification). The thermal upgrading content is also specified as 1.5 times the organic nitrogen content.

GE apparently shot for levels in the same ball park. They did have a Permalex II which had even higher nitrogen contents

Back as far back as 1958, GE proposed measuring the nitrogen content with infrared spectroscopy, but this was problematic with the state of the art of the instruments of that day. One could probably develop a good method using FTIR. There was a paper presented at Doble in 1997 by Cartlidge, Dominelli and Ashby from Powertech Labs and Polovick from BC Hydro who propose using

---

[1] Westinghouse Insuldur: A Proved Insulation System, Westinghouse Power Transformer Division, Sharon, PA, Publication SA 9025 B, 11-1967.

FTIR on methanol extracts of the paper to measure the dicyandiamide concentration.

Separate LTC Oil

Dissolved Gas Analysis (DGA) of oil in an operating transformer may be the single most useful diagnostic tool available to the transformer owner (see p.113). However, if you purchase a transformer with a Load Tap Changer (LTC) that shares oil in common with the transformer tank or that has a barrier between the LTC tank and the transformer main tank that is permeable to dissolved gases in oil, then you lose the majority (if not all) of the usefulness of this most important diagnostic tool. For instance, acetylene generated by a transformer is cause for immediate concern. A transformer producing acetylene could be in a failure mode and would require immediate investigation. LTC's, on the other hand, produce acetylene during normal operation. If the acetylene from the LTC masks acetylene produced by the transformer you could miss your opportunity to prevent a premature failure. Therefore, endeavor always to order transformers that keep the oil (and most importantly the dissolved gases in the oil) separate between the two compartments.

Basic Insulation Level (BIL)

Basic Insulation Level, or properly known as Basic Lightning Impulse Insulation Level, is described in IEEE C57 12.00-2000, *IEEE Standard General Requirements for Liquid Immersed Distribution, Power and Regulation Transformers.*

A trend that has become a practice is the building of higher voltage transformers with greater reduction in the BILs, or simply with less insulation. What is the rationale for building a transformer with lower BILs? One of the main reasons for lower BIL rating units is: cost – the transformer will have less insulation and that equals less money.

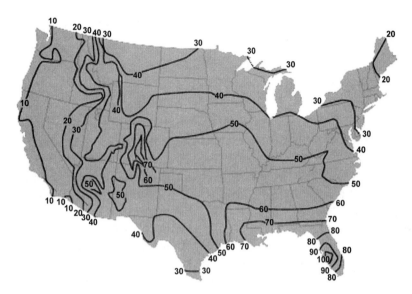

Figure 1.4 - Mean annual days of thunderstorm activity in the continental United States.

These practices are usually applied to units 69kV and above. Material costs such as solid insulation material, liquid insulation and steel, as well as transportation costs, can be realized. One of the requirements for safely using reduced BILs is the systems neutral be grounded. Reduced BILs are not available for use on ungrounded systems.

Are reduced BILs right for your situation? Consider the geographic location where the transformer will be located. Transformers located in areas where thunderstorm activity is high are at risk. For example, in the United States, the gulf coast region would not be a good candidate for reduced BILs. Parts of Florida can experience upwards of 100 days of thunderstorms a year. Conversely, the Pacific Northwest would see an average of 10 days of thunderstorms per year. That still doesn't mean you don't assume some risk. Several years ago the Pacific Northwest had a "freak" lightning storm with over 1000 documented lightning strikes in a one-hour period.

There are numerous web sites and other sources that monitor lightning activity, so evaluate carefully before choosing a unit with a lower BIL.

Standards

When purchasing a new transformer, what criteria do you follow? What information is available to you to help you make a decision during the purchase process? What electrical tests should be performed to ensure the transformer is built properly? These are tough questions.

Fortunately there are organizations out there such as The Institute of Electrical and Electronic Engineers (IEEE), the International Electrotechnical Commission (IEC), British Standards (BS), Canadian Standards Association (CSA), The Standards Association of Australia (AS), and the International Electrical Testing Association (NETA). These are the standards the majority of the world would follow in the purchasing, testing, installing, loading, and operating electrical equipment.

Most of us, especially in the United States use and follow the IEEE standards when purchasing/specifying the purchase of electrical equipment. In general, the standards recommend the minimum performance and design requirements. The IEEE also publishes guides to help in the testing, installing, and loading of transformers that have proven to be very helpful to the maintenance/utility engineer.

IEEE standards are developed by the various IEEE societies and the standards coordinating committees. Members of the committees serve voluntarily and without compensation. They are not necessarily members of the institute. The standards developed within IEEE represent a consensus of the broad expertise on the subject within the institute as well as those activities outside of IEEE that have expressed an interest in participating in the development of the standard.

For most electrical transformer applications, the *C57 Standards Collection* would be consulted. This collection in the past was published in its entirety and referred to as the "phone book." It was as big as a phone book with about sixty different standards, guides and recommended practices regarding distribution, power and regulating transformers. Unfortunately, the entire collection is only available on-line from IEEE at a yearly fee or on CD-ROM.

Some of the more popular standards available for consultation, and as described in the individual scope of the document are:

C57.12.00-2000 *IEEE Standard General Requirements for Liquid-Immersed Distribution, Power, and Regulating Transformers.* This standard was revised in 2000 and is a basis for the establishment of performance, limited electrical and mechanical interchangeability, and safety requirements for the described equipment; and for assistance in the proper selection of the equipment.

C57.12.90-1999 *IEEE Standard Test Code for Liquid-Immersed Distribution, Power, and Regulating Transformers.* This standard describes the methods for performing tests specified in C57.12.00 and other standards applicable to liquid-immersed distribution, power, and regulating transformers. It is intended for use as a basis for performance, safety, and the proper testing of such transformers.

C57.98-1992 *IEEE Guide for Transformer Impulse Tests.* This guide is written primarily for power transformers, but is also generally applicable to distribution and instrument transformers. Other standards already determine the specific requirements for impulse tests. The purpose of the guide is not to change these standards in anyway, but to add background information that will aid in the interpretation and application of these standards.

C57.91-1995 (currently under revision) *IEEE Guide for Loading Mineral-Oil-Immersed Transformers.* This guide cover general recommendations for loading 65 °C rise mineral-oil-immersed power transformers described in the standards.

C57.93-1995 (started revision) *IEEE Guide for Installation of Liquid-Immersed Power Transformers.* The recommendations presented in this guide apply to the shipping, handling inspection, installation, and maintenance of liquid-immersed power transformers.

C57.104-1991 (under revision) *IEEE Guide for the Interpretation of Gasses Generated in Oil-Immersed Transformers.* This guide applies to mineral-oil-immersed transformers and addresses:

    a) The theory of combustible gas generation in a transformer
    b) The interpretation of gas analysis
    c) Suggested operating procedures
    d) Various diagnostic techniques, such as key gasses, Dornenberg ratios, and Rogers ratios
    e) Instruments for detecting and determining the amount of combustible gases present
    f) A bibliography of related literature

C57.106-2002 *IEEE Guide for Acceptance and Maintenance of Insulating Oil in Equipment.* This guide applies to mineral oil used in transformers, load tap changers, voltage regulators, reactors, and circuit breakers. The guide discusses the following:

    a) Analytical tests and their significance for the evaluation of mineral insulating oil.
    b) The evaluation of new, unused mineral insulating oil before and after filling into equipment.

c) Methods of handling and storage of mineral insulating oil.
d) The evaluation of service-aged mineral insulating oil.
e) Health and environmental care procedures for mineral insulating oil.

Std. 62-1995 *IEEE Guide for Diagnostic Field Testing of Electric Power Apparatus – Part 1: Oil Filled Power Transformers, Regulators, and Reactors.* This guide describes diagnostic tests and measurements that are performed in the field on oil-immersed power transformers and regulators.

Std. 637-1985 *IEEE Guide for the Reclamation of Insulating Oil and for Its Use.* This guide covers mineral insulating oil commonly defined as transformer oil; definition and description of reclaiming procedures; the test methods used to evaluate the progress and end point of the reclamation process, and what criteria recommended for the use of reclaimed oils are considered suitable. This guide is currently being reaffirmed and then will be revised.

Some new guides that will be published soon are:

C57.130 2003 (draft 14) *IEEE trial-Use Guide for the Use of Dissolved Gas Analysis During Factory Temperature Rise Tests for the Evaluation of Oil-Immersed Transformers and Reactors.* This document provides guidance in the application of dissolved gas analysis (DGA) to transformers and reactors subjected to factory temperature rise tests.

C57.140-2003 (draft 9) *Guide for the Evaluation and Reconditioning of Liquid Immersed Power Transformers.* The goal of the guide is to assist the user in extending the useful life of a transformer.

For referenced IEEE standards, visit the IEEE website, www.ieee.org, or the bookstore at, www.sdmyers.com.

ANSI/IEEE and IEC Equivalents for Transformer Oil

In North American (and particularly in the United States), we use voluntary standards developed by two organizations – the Institute of Electrical and Electronic Engineers (IEEE) and the American Society for Testing and Materials (ASTM). There is a third organization, the American National Standards Institute (ANSI) that acts as a clearing house and central coordinating point for standards in many industries used in the United States. ASTM has recently changed its name to ASTM International and is pursuing a more global market for standards and standards development.

To summarize the roles of the three organizations, ASTM creates and publishes standards on test methods and material specifications. IEEE interprets the ASTM standards as they apply to the engineering needs of the electrical industry. ANSI reviews the ASTM and IEEE standards for their sufficiency in terms of clarity and completeness, specifically regarding how they meet the general needs of United States businesses and government agencies for adequate and understandable standards.

In the global economy, there is an international organization, the International Electrotechnical Commission (IEC). IEC basically assumes the roles of both ASTM and IEEE for the international community. IEC is made up of representatives from the national standards boards (ANSI is the United States' representative). IEC develops specifications and test methods, and provides their own interpretive guidance. Frequently, IEC adopts testing methods and other standards guidance from the International Organization for Standards (ISO). National standards bodies for various countries adopt IEC standards as their own (for example, the British Standards are published by a national organization in the United Kingdom). The use of Canadian standards in some cases is the primary exception to the general rule that IEEE/ASTM standards are used across North America.

IEC activities are supported by the technical actions of another international organization – CIGRE. The acronym is for the French translation of the organization's name: International Council of Large Electrical Systems. CIGRE is not in the standards "business," but technical work performed by members working on CIGRE task groups frequently becomes the basis for IEC standards.

The following paragraphs summarize the equivalent standards, and provides some brief commentary on the similarities and differences between North American standards of IEEE and ASTM and international standards of IEC and ISO. Although many of the standards cited below have applications to other fluids, we are going to concentrate only on the equivalent standards for mineral oil dielectric fluid.

New Oil Standards:

Organization	Standard No.	Title
ASTM	D 3487-00	Standard Specification for Mineral Insulating Oil Used in Electrical Equipment
ANSI/IEEE	C57.106-2002	IEEE Guide for Acceptance and Maintenance of Insulating Oil in Equipment
IEC	60296 Second Edition - 1982	Specification for Unused Mineral Insulating Oils for Transformers and Switchgear

Although it is not an "official" standard, a large number of transformer owners use Doble Transformer Oil Purchase Specifications (TOPS) as the basis for their internal new oil specifications.

Comment:

ASTM D 3487 lists two types. Type I is basically uninhibited oil (max. inhibitor content is 0.08%). Type II is inhibited (max. inhibitor content is 0.30%). IEEE C57.106 refers to ASTM D 3487 as the appropriate specification for oil at the refinery and provides further standards for oil as received from the refinery, new oil received in new equipment (broken down by voltage class), new oil processed for 345 kV class equipment, prior to filling, and new oil in equipment, after processing, filling, and standing, immediately prior to energizing.

Purchase | 31

The IEC standard is considerably different. There are three classes of uninhibited oil, classified by different ranges for viscosity, flash point, and pour point. There are also three classes for inhibited oil. None of the three classes in either case correspond closely to ASTM specification oil. Note the key differences in the table below, concerning physical properties related to flash point, viscosity, and pour point.

	ASTM	IEC Class I	IEC Class II	IEC Class III
Viscosity at 40 °C, cSt, maximum	12	16.5	11	3.5
Flash point, minimum	145 °C	140 °C	130 °C	95 °C
Pour point, maximum	-40 °C	-30 °C	-45 °C	-60 °C

Oil manufactured according to the IEC specification may not perform as per design in an IEEE standard transformer. This is a concern for foreign manufactured transformers being used in the United States.

Note: At press time, IEC 60296, third edition had been approved. Time was not sufficient to present the change in this edition.

In-service Oil Standards:

Organization	Standard No.	Title
ANSI/IEEE	IEEE Standard 637-1985	IEEE Guide for the Reclamation of Insulating Oil and Criteria for Its Use
ANSI/IEEE	C57.106-2002	IEEE Guide for Acceptance and Maintenance of Insulating Oil in Equipment
ANSI/IEEE	IEEE Standard 62-1995	IEEE Guide for Diagnostic Field Test of Electrical PowerApparatus – Part 1: Oil Filled Power Transformers, Regulators, and Reactors
IEC	60422 Second Edition, 1989	IEC Supervision and Maintenance Guide for Mineral Insulating Oils in Electrical Equipment

Comment:
See our white paper comparing IEEE/IEC standards for in-service oil to the oil classification system used by S. D. Myers. Limits for in-service oil in the IEEE and IEC standards are not sufficient to maintain a cost effect maintenance program to maximize the reliable life of the equipment.

The basic limits established for IEEE standards have, to date, been based on the limits contained in Standard 637. These have not changed since the standard was first published in 1985. There are some important changes incorporated into C57.106-2002.

The IEC document is a replacement for the original Standard 422, published in 1973. This standard is also undergoing revision.

Standard Testing Methods
Table 1 lists the methods used in the oil standards for both ASTM and international organizations. Some of these, as noted, cannot be considered direct substitutes.

1. IEC 60296 contains descriptions and modifications of methods as well as references to ISO methods. Where the only reference is to an ISO method, this discussion does not refer to 60296.

2. The standard in the United States is to use the Cleveland open cup for transformer oil (ASTM D92). The international standard is to use the Pensky-Martens closed cup (ISO 2719).

3. IFT is reported in units of dynes/cm (US) or mN/m (standard metric). These are equivalent – 1 dyne/cm is equal to 1 mN/m.

4. Although the three dielectric methods all report breakdown strength in kV, the measurements are not equivalent. D 877 uses flat disk electrodes, spaced 0.10 inches apart. D 1816 uses VDE (spherical) electrodes, spaced either 1 mm (0.04

inches) or 2 mm (0.08 inches) apart. IEC 60156 uses VDE electrodes spaced 2.5 mm apart.

5. Liquid power factor (ASTM) and tan δ (IEC) are not exactly the same measurement, although they correspond fairly closely in the operating range of most U.S. transformers. This must be considered since most foreign operating transformers operate outside this range where the two measurements are roughly equivalent. Also, for the ASTM method, the determination is made at 25 °C and 100 °C. The IEC standard temperatures are 20 °C and 90 °C. Liquid power factor is typically reported as a percentage (i.e., 0.10%) while tan δ is typically reported as a decimal number (i.e., 0.001).

6. Resistivity is not used to a great extent in North American standards for in-service oil. We included it in the chart because the results between the ASTM method and the IEC method are typically equivalent, so a more direct comparison is possible than using liquid power factor compared to tan δ.

7. This is the infrared spectrophotometer method. There is an ASTM gas chromatography method (D 4768), but no corresponding IEC GC method.

8. The methods are not equivalent. The new oil specifications and oil standards should be compared to the standard methods when determining which of the methods to run.

9. Packed column chromatography with matching of Aroclor pattern is the standard for the U.S. (D 4059). This is roughly equivalent to IEC 60997, which is no longer used. The IEC standard is for capillary column chromatography, with quantification of individual congeners. There is no directly equivalent ASTM (or even EPA) method.

10. The U. S. standard for interpreting dissolved gas analysis is ANSI/IEEE Standard C57.104-1991, Guide for the Interpretation of Gases Generated in Oil-Immersed Transform-

Property Tested or Standard Guide	ASTM	IEC	ISO
Guide to Sampling (General)	D 923	60475	
Guide to Sampling (DGA)	D 3613	60567	
Color and Appearance	D 1500 D 1524	60296[1]	2049
Flash Point (open cup)[2]	D 92	60184	259
Flash Point (closed cup)[2]	D 93	60034	2719
Hydrocarbon Type	D 2140	60590	
Interfacial Tension[3]	D 971	60296	6295
Pour Point	D 97		3016
Specific Gravity (Relative Density)	D 1298	60296	3675
Kinematic Viscosity	D445		3104
Dielectric Breakdown Strength[4]	D 877 D 1816	60156	
Power Factor, Dissipation Factor[5]	D 924	60247	
Resistivity[6]	D 1169	60247	
Acid number (neutralization number)	D 974	60296	
Corrosive Sulfur	D 1275		5662
Inhibitor[7]	D 2668	60666	
Oxidation Stability[8]	D 2112 D 2440	60074 60474 60813 61125	
PCB content[9]	D 4059	60997 (packed) 61619 (capillary)	
Moisture	D 1533	60814	R760
Furans	D 5837	61198	
Dissolved gas analysis[10]	D 3612	60567	

Table 1

ers. The corresponding IEC standard is IEC 60599, Mineral Oil-Impregnated Electrical Equipment Inservice – Guide to the Interpretation of Dissolved and Free Gases Analysis. The IEC method for analysis (IEC 60597) also includes the sampling standard. For ASTM, there is a separate sampling standard – ASTM D 3613.

Factory Electrical Tests

Factory electrical tests are performed to verify that transformers meet design and purchase specifications and also provide an initial set of benchmarks to be used for comparison purposes against future field electrical tests.

IEEE Standard General Requirements for Liquid-Immersed Distribution, Power, and Regulating Transformers C57.12.00-2000 defines the factory tests in three categories: routine, design and other. The tests are described as follows:

Routine
Tests made for quality control by the manufacturer on every device or representative samples, or on parts or materials as required, to verify during production that the product meets design specifications.

Design
Tests made to determine the adequacy of the design of a particular type, style, or model of equipment or its component parts meet its assigned ratings and to operate satisfactorily.

Other
Tests so identified in individual product standards, which may be specified by the purchaser in addition to routine and design tests. Examples: impulse, insulation power factor (class I only as it is a routine test for class II), and audible sound.

The following list of tests would typically be included in the factory tests.

- Resistance measurements: these are sometimes referred to as winding resistance tests because the integrity of the winding is being checked. Poor results can indicate problems with the winding or connections under tests.

- Ratio test (turns ratio): This test would give an indication if the transformer has been constructed with the proper ratio of turns in the primary and secondary windings and if the transformer has the proper percent differential between taps.

  The test should be performed on all rated voltage connections on all tap positions. +/- 0.5% from the calculated nameplate value would be the standard deviation that would be acceptable.

- Polarity/phase relation is performed to prove that windings are connected per the nameplate.

- No load loss and excitation current at rated voltage: These losses are primarily core losses. Dielectric & copper losses are minor contributors to no-load losses.

- Temperature rise test: Per the standard, this test may be omitted if data is available on a duplicate unit for certain voltage/kVA classes. The factory temperature rise tests are to determine weather the temperature rise of the windings, oil, and other components meet the design values.

  - Dissolved gas analysis during the factory temperature rise test is described in *IEEE Trial-Use Guide for the Use of Dissolved Gas Analysis During Factory Temperature Rise Test, C57.130*. The standard has criteria for the acceptable, possible problem and certain problem levels of the hydrogen component, hydrocarbon component and levels of the carbon oxides gasses.

The dielectric tests that are to be performed include:

- **Applied potential test** - what you may refer to as a hi-pot or over-potential test.

  The major insulation is tested at low frequency voltage (60hz) without exciting the core (1-min). This is a go-no-go test and the standard; *IEEE Standard Test Code for Liquid-Immersed Distribution, Power, and Regulating Transformers* C57.12.90 describes the failure mode as smoke or a rise in the leakage current.

- **Impulse tests** - These tests are to simulate something the transformer may experience during its service life many times over.

  - One Reduced Full Wave (establishes wave pattern)
  - Chopped Wave (simulates voltage collapse or traveling wave flashing over insulator)
  - Full Waves (simulates distant lightning stroke)
  - Front of Wave (simulates direct hit of lighting)
  - Switching impulse test simulates switching by the utility

  These tests are performed at percentages of the transformers rated basic lightning impulse insulation level (BIL).

- **Induced potential test** - Testing the major & minor insulation (turn to turn & coil to coil) at a higher frequency (120/400hz) to avoid excessive phases to phase voltages. This would be the last dielectric test performed.

  Induced potential test for distribution and class I power transformers: the test voltage varies with BIL level and the duration of the test is 7200 cycles. Look for smoke and leakage current as a failure mode during this test.

Induced potential test for class II power transformers is a bit different. The test voltage is based on system voltage. The duration of the test is 7200 cycles at the enhanced level followed with a 1-hour test to record any RIV activity. The RIV is measured at bushing test tap and recorded at 5-minuite intervals for 1-hour. The maximum RIV is not to exceed 100 microvolts and the RIV rise during 1-hour test not to exceed 30 microvolts.

The insulation power factor test is recommended at the factory for two reasons: (1) it will give a good indication of the insulation dryness, and (2) it will give a benchmark for future testing. We have good correlation with the power factor and the percent moisture in the insulation by weight in the areas of 2% and less. Therefore if you have a 0.5% power factor, the moisture content of the insulation is approximately the same, equaling a properly dried insulation system. Also specify the desired test method with or without the guard circuit (I or II). Method II is more representative of field test. (C57.12.90)

It may be wise to specify dissolved gas tests after filling, after temperature rise test (if required) and after induced potential tests.

Audible sound testing may be specified as well depending on the transformers use. If required to be in service in a residential area where the noise may be of concern, testing should be specified. See standard C57.12.90 for the test procedures.

In the Annex to Part I of C57.12.90, the minimum information to be included in a certified test report is described. It is recommended this and all other information pertaining to a specific transformer, be properly maintained for reference in the future.

Site Installation

When the transformer shows up on site, what will you do? It starts with proper planning. Installation instructions are a must! You need to create them, or, review the installation instructions of the company hired to perform the work. These would include, but are not limited to: procedures for testing, moving, vacuum filling, oil processing and commissioning the unit. Many times we have received invitations to bid on jobs of this sort and the specification just calls for "install the unit" or "fill the unit." Remember, you just might get what you ask for! However, most, if not all of the electrical testing and oil qualities are described in the ANSI/IEEE standards. These standards are dependent upon the voltage class, BIL rating, etc. Specific instructions for installation must be included as part of the job contract to prevent misunderstandings and identify responsible parties.

Specifications

Review the specifications with your proposed installation contractor so everyone knows what is required. You may have your own specifications, and your contractor may have theirs, as well. Compromise may be in order, and a portion or the best parts of each may be used. The *IEEE Guide for Installation of Liquid-Immersed Power Transformers C57.93-1995* provides an industry consensus standard and is available for consultation. In any case, the manufacturer's specifications will most likely supercede any other specifications because of the warranty.

The more you can cover in the specifications, the better the job is defined, and the more accurate the true costs can be estimated. We have seen specifications from "install the unit" to specifications that will detail what form is required to report the results of the inspection and testing, and just about everything in between.

Site Preparation

The site for the installation of a transformer must be carefully considered. A level foundation designed for the weight of the transformer should be constructed. Select a location where air can circulate freely as we do not want to disrupt the designed cooling capacity.

Prior to arrival of the new or remanufactured unit, the following should be considered:

- What is the weight of the transformer?
- Is the existing or new pad adequate to support the weight?
- Is the existing or new oil containment adequate for the amount of oil?
- Are the road surfaces adequate to support the weight? Remember to include transport vehicles and rigging equipment when making this determination.
- Are there any buried water lines, sewers, or conduits that may collapse under the added load of the transformer?

Shipping

The method of shipment depends on the class, size, shipping clearances required, weight restrictions, and facilities for installation. Transformers are bolted, chained, braced and rodded for either rail or truck shipment. Whenever it is practical, transformers are shipped by truck. This method provides a "softer" ride than rail shipment. Therefore, there is less chance of damage due to impact forces.

Transformer Arrives by Truck

- Did the transformer arrive with positive pressure?
- Is there any evidence of external damage?
- Take the dew point of transformer.

- Perform acceptance testing if applicable.
- Relieve pressure on the transformer.
- Perform air quality checks. Purge with dry air. IEEE Standard C57.93 recommends a dew point of -62 °F at 30 °F ambient temperature for purging air.
- Perform an internal inspection. Before entering the unit, verify the internal atmosphere is non-hazardous and with an oxygen content of at least 19.5%, in compliance with OSHA requirements.
- Care should be taken to limit the exposure time to atmosphere.
- Note any discrepancies or signs of damage.
- Is there any reason to notify the manufacturer?
- Does a claim against carrier need to be filed?

Transformer Arrives by Rail

- Did the transformer arrive with positive pressure?
- Is there any evidence of external damage?
- Examine impact recorder tape(s) if used. This official document must be safeguarded at all times. It can and frequently is used in a court of law to verify mechanical damage during transport.
  - Note any impacts that register in the 3 G or higher category.
  - If any 4 G impacts have occurred, notify factory representative as soon as possible.
  - File claim with railroad for hidden internal damage when impacts of 3 G or higher occur. The exact amount of gravitation forces that a specific transformer can endure is based on its design. The manufacturer must provide this information.
- Take the dew point of the transformer.
- Perform electrical acceptance testing as per IEEE/ANSI C57 Standards, manufacturer's specifications, and good engineering practices.

- Relieve pressure on the transformer.
- Perform air quality checks. Purge with dry air. IEEE recommends a dew point of -62 °F at 30 °F ambient temperature for purging air.
- Perform an internal inspection. Before entering the unit, verify the internal atmosphere is non-hazardous and with an oxygen content of at least 19.5%, in compliance with OSHA requirements.
- Care should be taken to limit the exposure time to atmosphere.
- Is there any shipping damage or irregularities?
- Is there any reason to notify the manufacturer?
- Does a claim against the carrier need to be filed?

Internal Inspection

All confined space/permit entry requirements are to be followed while performing internal inspection.

During the internal inspection, it is recommended the transformer be continuously purged with dry air and monitored for air quality.

- Check to be sure all internal hardware is tight.
- Rotate the de-energized tap changer and inspect contact alignment.
- Inspect windings:
    - Does there appear to be any movement of windings or supporting members?
    - Are spacer columns straight?
    - Is the blocking tight?
    - Are the current transformer leads tied away from the phase conductors?
- Check corners of core for misalignment or any other sign of core movement or bent core steel at the laminations.

Unloading and Transporting

- What equipment is required to match the weight and size of transformer for lifting and moving?
- Are the roads or surfaces adequate to support the weight of vehicles and equipment plus the transformer?
- Is the selected rigger familiar with handling transformers?
- Are the equipment and rigging materials free of defects and adequate for the demands to put upon them?
- How far does the transformer have to be moved and are there restrictions for travel?

Field Assembly (Dress Out the Unit)

Transformers with accessories and equipment removed for shipment must be reassembled after being placed on the installation site. Some very tall transformers will have temporary shipping covers that must be removed and the permanent covers installed. Field assembly is a project that needs careful coordination by the project manager.

Some things to remember:

- The assembly process should be planned so that the internal parts of transformer have minimal exposure to atmosphere.
- Before this phase starts it would be a good time to do an inspection and inventory of the parts and accessories needed.
- Rigging will generally be needed to mount the bushings, arresters, conservator tank, electrical junction boxes, oil pumps, radiators, fans and other devices too heavy to lift by hand.
- Uncrate and clean bushings prior to opening the transformer.

- The manufacturer will generally supply all gaskets and seals required to assemble the new or remanufactured unit. Do not reuse gaskets and seals!
- While performing any internal work during the assembly, this would be a good time to remove any special bracing in the transformer that was installed for shipping purposes, or confirm the removal of such bracing if it was performed when the transformer arrived. This bracing is usually color coded and clearly marked to be removed after shipping and before installation.
- Log any tools and hardware that are taken into transformer.
- Account for any tools and extra hardware after bushings are installed and transformer is ready to seal up.
- After assembly has been completed, pressure test the unit with dry air or nitrogen and conduct an eight (8) hour (overnight) leak test or follow manufacturer's instructions for leak test.

Vacuum Leak Check the Transformer

No matter how well the transformer performed during the leak test under positive pressure, another leak test must be performed under vacuum to insure that the unit is ready for the vacuum treatment. It will do absolutely no good to start a vacuum treatment if the transformer has a leak and you are pulling outside air into the tank.

Pull vacuum on the transformer and perform a leak rate test by holding a vacuum of 2000 microns (2.0 torr) for 30 minutes on the transformer tank. Record the transformer vacuum gauge reading, and then close the vacuum valve on top of the transformer. Record the vacuum readings periodically for 30 minutes. Compare with allowable leak rate chart.

Excessive vacuum pressure losses at this vacuum level (2000 microns) would indicate leaks that must be repaired before vacuum filling can proceed. This may involve the use of sonic detectors to locate the leak, duct-sealing compound for sealing small leaks, or welding to repair larger leaks. Leak detection sometimes can be very difficult under vacuum conditions. You may need to perform another leak test under positive pressure to find the source of stubborn leaks.

Vacuum Filling the Transformer

Follow the manufacturer's instructions for vacuum levels, moisture limits and oil filling rate for warranty considerations. This step is one of the most important steps in the installation process. Failure to complete these steps successfully may result in starting the vacuum treatment again or possibly commissioning a unit not properly processed that then could lead to accelerated aging of the insulation system and valuable life loss. If this would go undetected, it would lead to early failure of the transformer.

- What was the initial dew point and what dew point do you need to achieve?
- The ultimate goal is to have a dew point correlate to 0.5% or less of moisture content.
- If a good dew point was initially achieved and the exposure time to transformer was limited, vacuum may be the only requirement prior to final filling of the transformer. A good rule of thumb can be six hours of vacuum under 500 microns for each hour of exposure.
- If a good dew point was not achieved or if exposure time was greater than desired, a hot oil and vacuum cycle will be required. This process is also recommended for any transformer rated 230 kV or higher.
- Cover the core and coil of transformer with transformer oil. Circulate transformer oil through heater. Vent transformer or pull a small amount of vacuum. Most processing rigs are not equipped to circulate oil under vacuum.

- Record oil temperature going into and coming out of transformer. Limit temperature of in-going oil to a maximum of 90 °C.
- Circulate oil through transformer until bottom oil temperature reaches 60 °C. If there are cold weather temperatures when this process is taking place, it is recommended that the transformer be blanketed with some type of insulated tarp to prevent excessive heat loss.
- When the desired bottom oil temperature has been reached, drain transformer and immediately apply vacuum to transformer.
- Maintain vacuum < 500 microns for 24 hours.
- Periodically record the transformer vapor pressure; valveing off the vacuum and monitoring the pressure loss can obtain measurements of the vacuum loss rate. This is no longer a sign of leaks but a sign that moisture is still being removed from the insulation and the moisture is displacing pressure.
- When there are no longer signs of pressure loss, the final measurement of vapor pressure along with the temperature can be used to plot the insulations moisture content on the Piper Chart. (Figure 7.2, p. 279)
- At the end of vacuum cycle, relieve vacuum with dry air and bring transformer to a positive pressure.
- Take a dew point reading after the dry air has been on the transformer for 24 hours.
    - If the dew point reading is good, pull an addition eight hours of vacuum under 500 microns and final fill transformer. The heat and vacuum cycle may be repeated if the desired results are not achieved.
- Transformer should be filled with transformer oil that has been tested and meets IEEE C57.106-2002 *Guide for Acceptance and Maintenance of Insulating Oil in Equipment Standards*. Follow the manufactures recommendations on fill rates.
- The vacuum level on the transformer should be maintained at a maximum of 1000 microns during filling.

- Once transformer has been final filled, apply a positive pressure with dry air or nitrogen and search thoroughly for any sign of leaks.

Any final oil and electrical testing should be completed during and after the units recommended set time, as established by the work scope of the project.

There are many different procedures to choose from. Some specifications are the bare minimum and some are overkill. The information included in this book is intended to help you in making these decisions. Remember, it would be better to err on the safe side. Transformers are expensive and not easily replaced. If the site installation is properly completed, and future maintenance issues are properly addressed, transformers can be expected to have a long life.

Chapter 2 | Warranty

As a transformer owner (or as someone responsible to the transformer's owner), you can't take a vacation from responsibility during a new transformer's warranty period. While it is true that the manufacturer will take responsibility for defects in materials or workmanship during the warranty period, not all such defects manifest themselves during the warranty period.

One-Month DGA

If a new transformer, when energized, survives the first ten cycles, then the first ten seconds, then the first ten minutes, chances are pretty good that it will also survive the first ten years. However, during the warranty period the transformer could have a fault that wouldn't be manifest during the first ten minutes of operation – whether from manufacture, transportation or installation – that could lead to failure. Most often, such a fault will leave its signature of dissolved gases in the oil.

Therefore, we recommend that you run an initial dissolved gas analysis (DGA) on a newly installed transformer around the first full month in service. This test will indicate any of several faults that could potentially lead to failure. Even though the manufacturer would pay to repair or replace a failed unit under warranty, it would be in everyone's best interest to identify a fault before it leads to failure.

Ten-Month Full-Oil Test

Some defects that will not become obvious until after the warranty period will nonetheless be observable in oil test results, particularly in the DGA.

Since every responsible transformer owner will at some point begin annual oil testing, we strongly recommend that you begin a new transformer's annual oil testing about two months before the warranty expires. This will give you time to have the sample analyzed, review the test results and take up with the manufacturer any abnormalities you might find – before the warranty expires. If the test results don't indicate any problems, you now have established the baseline for that transformer's oil test results. You can compare future test results to these first year results to begin plotting trends throughout the transformer's life.

Chapter 3
Oil Testing

Introduction to Transformer Oil

Why do transformers have oil in them? As we examine this topic further we will find that life might be a lot simpler if they did not. We will review the effects that the aging of the oil has on transformers and consider the inconvenience and expense of testing and maintaining the oil. However, we will also find that oil contributes substantially to the operation of transformers. We could not operate our power systems without it. So, how did we come to depend upon it so much?

Mineral Insulating Oil

First of all, let's make sure that we understand and can properly define what we are talking about. For that, we are going to take our definitions from an ASTM standard, D 2864-02 Standard Terminology Relating to Electrical Insulating Liquids and Gases, which is under the jurisdiction of ASTM Committee D 27 on Electrical Insulating Liquids and Gases. Here are some terms:

Dielectric – A medium in which is possible to maintain an electric field with little supply of energy from outside sources. The energy required to produce the electric field is recoverable in whole or in part. A vacuum, as well as any insulating material, is a dielectric.

Insulating material – A material of relatively low electrical conductivity and high dielectric strength, usually used to support or provide electrical separation for conductors.

Insulating liquid, fluid, or gas – A fluid (liquid or gaseous) which does not readily conduct electricity. Electrical insulating fluids typically provide both electrical insulation and heat transfer in electrical equipment.

Dielectrics and insulating materials can be solids, liquids, or gases – and even vacuum can be a dielectric. An insulating liquid is a type of insulating fluid. Further, it is both an insulating material and a dielectric. All of that is pretty straightforward, but things can get confusing when terms get misused. Unfortunately, some people refer to anything other than air inside a transformer as the "oil" in the transformer. So, when they say "transformer oil," they may really mean "insulating fluid" or "insulating liquid." We are going to tighten up on the definition further to prevent confusion. We are going to use "transformer oil" or "oil" only when that is what we mean – a substance specifically called "mineral insulating oil." So, we better define that term:

> **Mineral insulating oil** – An oil of mineral origin, refined from petroleum crude oil, possessing electrical insulating properties.

This is what we mean throughout this book when we talk about "transformer oil." Similar terms you may see or hear, from us or from other sources, that also mean the same thing are: mineral oil dielectric fluid (and its acronym – MODEF), petroleum based insulating oil, transformer mineral oil, and mineral oil. There are other insulating fluids, liquids and gases, that are used as dielectrics in transformers, as well.

So, when we talk about oil, we are talking about mineral insulating oil, widely used in transformers with Kraft paper solid insulation. Let's answer the first of the two questions we asked earlier. Why do transformers have oil in them?

Ancient History of Transformer Oil

The very first transformers did not have oil in them. The only fluid insulating medium or dielectric was air. There was a severe problem with this because of the materials of construction that they had to work with. The early designs had relatively large energy losses that generated a lot of heat in the core and coils. The air was not completely effective at transporting that heat away from the center of the transformer. Size was limited to very low capacity designs – any bigger and the inefficient devices would literally burn up.

This situation limited the usefulness of the transformer, and it was apparent almost immediately that some innovation was needed to overcome the problem. Elihu Thomson patented the use of mineral oil in transformers in 1887, and, by about five years later, the practice of using oil as a dielectric was commercially established.

The "Four Functions"

Early designs depended on mineral insulating oil to provide heat transfer and to keep the energized parts insulated from each other. As use of mineral oil in particular and insulating fluids in general has developed, and, as our technology and understanding concerning them has expanded, it has become clear that the mineral oil in an oil filled transformer fulfills four functions that contribute to the operation of the transformer.

The four functions of transformer oil are:

1. Oil provides dielectric strength – acts as a dielectric and insulating material.

2. Oil provides heat transfer – acts as a cooling medium.

3. Oil protects the solid insulation – acts as a barrier between the paper and the damaging effects of oxygen and moisture.

4. Oil can be tested to give an indication of conditions inside the equipment – acts as a diagnostic tool for evaluating the solid insulation.

As it ages, the oil's ability to fulfill some of these functions is compromised. Chemical by-products from aging build up in the oil, and even fall out or precipitate from the oil as sludge. If aging is allowed to proceed long enough, this will affect the oil's capacity to act as a dielectric or to effectively transfer heat. Oil needs to be beat up pretty badly in service in order for these effects to become noticeable. Further, with regard to the fourth function, oil in almost any condition can be tested and the results provide some useful information. Regardless of how much the oil has aged, that function is only compromised to a minor degree, if at all.

The function of oil that is most important from the perspective of oil and equipment maintenance – protecting the paper – is also the most sensitive to the effects of the oil aging in service. The decay products that build up over time, and that have to advance significantly in order to affect dielectric strength or heat transfer capabilities, begin to affect the oil's protective qualities almost immediately after they start to be formed. Very small quantities of these decay products form early in the aging process. These have a high affinity for the structure of the paper that makes up the solid insulation, and aggressively tear the molecular structure of the paper apart when present in even extremely small concentrations. This means that the function of oil that is most important is the function that aging takes away first.

Our oil maintenance program is going to be defined by the necessity of keeping these aging decay products of the oil away from the paper. The way that we do that, as we will discover later in this discussion, is to prevent the formation of these decay products to the greatest degree possible and to remove them from the oil and especially the paper as soon as possible if they do form.

FUNCTION 1 - Oil as a Dielectric

The two materials that are used most extensively as insulation in transformers are Kraft paper (as the solid insulation) and mineral oil (as the insulating liquid). The most important insulating medium – or dielectric – in the equipment is actually neither the liquid nor the solid. Rather, it is the *combination* of both materials. There is a synergy between the two materials when it comes to dielectric strength. The combination of oil soaked paper is a stronger dielectric than would be expected if the dielectric strength of the oil was added arithmetically to the dielectric strength of the paper. A pretty good estimate is that the combination of Kraft paper and mineral oil dielectric fluid has a dielectric strength that is about 20 to 25% greater than one would expect from just adding the dielectric breakdown strength of the two materials together.

In addition, the dielectric strength per pound and per volume is impressive. There are a number of materials that are similar in electrical capabilities, but none are as versatile for the price. The paper/oil combination was introduced well over 100 years ago – and no one has found a cost effective replacement for most of the applications where this system is used.

Aging in service affects the dielectric properties of the oil (as a stand alone medium) more rapidly than it does similar properties for the solid insulation on the same basis – but the change in the dielectric properties of the oil upon aging is a very slow and gradual deterioration. At a certain critical point, as oil aging products fall out of the system as sludge, the gradual decline in dielectric properties of the oil can accelerate rapidly as conductive paths in the oil are formed, bridging charged conductors in the equipment.

Aging does affect the properties of the oil/paper combination more rapidly than is the case for either stand alone material, but even those changes in the *electrical* properties take place much more slowly than aging affects the physical and chemical properties of the two materials.

FUNCTION 2 - Oil as a Heat Transfer Medium

The physical properties of the oil are determined by the composition of the oil. The distribution and type of hydrocarbons present, and the molecular weight of those components, determine key physical properties such as the viscosity profile, specific heat, relative density, and coefficient of expansion for the oil. These key physical properties determine how well the oil will move to a point where heat can be "picked up" from the core and coils, absorb the excess heat, transport the excess heat from where it is not wanted to the shell of the transformer, and finally dissipate the excess heat to the atmosphere.

As it ages in service, the oil's characteristics with regard to heat transfer will change very slowly. For a long period of time, until solid sludge builds up in the oil, there will be basically no change. As solid sludge forms, it will start to precipitate out of the oil. This will impede heat transfer by coating the metal surfaces where heat is dissipated by the oil to the atmosphere. Sludge from aged oil will also fill in pores and other structures in the solid insulation preventing complete passage of the oil into and out of the coil. That condition further impedes proper heat transfer. The physical properties that define the oil's heat transfer capacity do not change substantially during the aging properties, so almost all of the effects concerning heat transfer are due to sludge formation.

FUNCTION 3 - Oil as a "Paper Protector"
A transformer is, quite literally, held together by the paper that forms the solid insulation. The effects of heat, oxygen, and moisture serve to break the paper down, decreasing its mechanical strength. 85% of all transformer failures occur because the paper has weakened to the point where it can no longer recover from stress. The paper (and the transformer) fail when the weakened paper gives way instead of "bouncing back."

Oil protects the paper from the effects of heat, oxygen, and moisture. The oil carries heat away from the paper in the coil and core as it is formed. Properly installed transformer oil has a much lower level of both oxygen and moisture content than the atmosphere does, so it forms an effective barrier to the detrimental effects of oxygen and moisture.

As the oil ages, reaction products form in the oil and paper. These reaction products are very aggressive toward the paper and actively tear it apart, molecule by molecule. This dramatically reduces the mechanical strength of the solid insulation. This reduction in strength begins to happen immediately to a significant degree as soon as any by-products form from the aging of the oil. Since this affects the oil's capacity to fulfill its third function, it is clear that that function of the oil is the first to "go" upon aging. Oil maintenance restores the oil ability to protect the paper because it removes the detrimental decay products.

FUNCTION 4 - Oil as a Diagnostic Tool

Through several decades of oil testing and careful correlation of test results to conditions inside electrical equipment, we have learned to interpret the results from oil tests. Although aging of the oil affects the test results, aging does not limit our abilities to either test the oil or interpret the results. We can always get a value for the oil tests we run to evaluate and monitor conditions in the equipment. Further, we can always apply a valid interpretation to those results concerning conditions inside the equipment. Aging of the oil does not affect whether the oil can fulfill its fourth function.

What Causes the Oil to Age?

The previous discussion talked about what happens when oil ages in service. We next need to define what we mean by that phrase "ages in service." Passage of time is only a small part of the answer. Oil that is many years old can still be relatively "new" in terms of its beneficial characteristics, while oil that is only a few years old can be so "aged" that it requires substantial clean-up and maintenance to restore its functionality.

Oil ages because it oxidizes. The hydrocarbons in the oil react with dissolved oxygen to form oxidation byproducts in the oil. These oxidation byproducts are what we mean when we use terms such as decay products, oxidation products, oxidation compounds, or aging by-products in the oil. Oil ages and oxidizes at different rates depending on a number of variables. A faster rate of aging means that the oil requires more frequent maintenance – and also that there is more damage done to the solid insulation if the required maintenance is delayed.

The reaction of transformer oil with oxygen produces oxidation products that fall into one of the general families of chemical compounds, depicted below with a generic structure for each family of compounds:

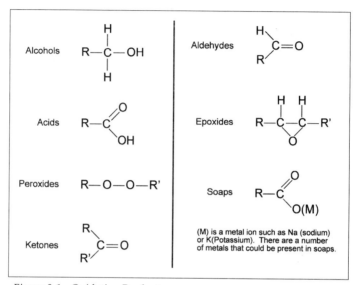

Figure 3.1 - Oxidation Products

These chemical compounds have several characteristics in common:

1. All of these compounds contain a hydrocarbon chain – "R" is a designation used by chemists as shorthand to denote a hydrocarbon chain. (The major components of transformer oil are hydrocarbon chains of different types, lengths, and molecular weights.)
2. All of the oxidation products also contain oxygen. Further, the oxygen is contained in a part of the molecule known as a functional group.
3. These functional groups containing oxygen cause the compounds to be "polar compounds." Polar compounds have a number of chemical characteristics

in common, even when they are considered to be of different "families" of compounds. Some of these common characteristics affect how we deal with these oxidation products in aging transformers.

4. Because they are polar, these oxidation products can be removed from the oil by a maintenance procedure that passes the oil through a bed of an adsorbent material such as fullers earth.
5. Also because of the polar content, these compounds are partially soluble in oil and partially soluble in water. A side effect of this characteristic is that service aged oil will attract and hold more moisture than new oil.
6. All of these oxidation products are aggressive toward the paper that makes up the solid insulation. They chemically attack the cellulose chains that make up the paper fibers.

Further, these decay products act upon each other and upon the oil to form sludge. Sludge is less soluble in oil and falls out of solution, forming deposits in the equipment and interfering with heat transfer and dielectric strength, as discussed above. Because of their polar character, these decay products form first within and on the solid insulation – cellulose also has polar characteristics and has a greater affinity for these decay products than the hydrocarbons making up the oil, which are not polar.

So, when we refer to aging of the oil in service, what we are really referring to is the oxidation of the oil molecules caused by the fact that there is oxygen dissolved in the oil.

Accelerating the Rate of Aging

What makes oil age faster? Another way to ask this is what increases the rate of oil oxidation. It turns out that there are a number of conditions in a transformer that will make the reaction of oil and oxygen go faster. Let's look at what affects the speed at which oil oxidation proceeds.

Oxygen Content

It may seem a little obvious, but it is still worthwhile to point out that oxygen content has a direct impact on how fast the oxidation reaction of transformer oil proceeds. Below a certain level, there is not enough oxygen present to sustain any measurable degree of oxidation. We consider 1000 to 2000 parts per million of dissolved oxygen to be the lower limit necessary to support the oxidation of transformer oil.

The more oxygen, the faster the reaction of oil and oxygen, and it is very close to being a direct relationship. Some care needs to be taken, though, when considering oxygen content. In addition to the oxygen dissolved in the oil, oxygen is also present in the oil and in the paper, chemically bound to the molecules of oil and paper. At higher temperatures above 70 °C, that oxygen, which is normally not available to oxidize the oil, may be released to react with oil and to accelerate the aging process.

Most of the air present in transformer oil is atmospheric in origin. Air contains about 20% oxygen, and the oxygen dissolves in oil exposed to the air. When the oil is installed, the oxygen is removed by vacuum degassing. A transformer that is completely open to the atmosphere (a "free breathing" transformer) would have about 30,000 ppm of oxygen dissolved in the oil. A sealed unit, where the oil was vacuum degassed, with a nitrogen blanket, will have about 90% of the oxygen removed, down to about 3,000 ppm. A transformer that is vacuum filled and that has a good oil preservation system such as a diaphragm conservator or a continuous pressure demand nitrogen system will have about 99% of the oxygen removed, down to about 300 ppm.

Oxidation is not a major concern if the oil is less than about 1000 to 2000 ppm of dissolved oxygen. For oil that exceeds this level, oxidation can be further reduced by using an oxidation inhibitor. Oxygen content of the oil will increase as the equipment is in service due to leaks and to oxygen released from chemical bonds in the oil and paper. Oxygen is typically removed from the oil at installation. Oxygen can also be removed from a transformer that has been in service by vacuum degassing.

Heat

Many chemical reactions proceed faster at higher temperatures. The oil oxidation reaction is one of these. The study of reaction rates in chemical reactions is a well-known specialty in physical chemistry called "kinetics." So many reactions show a similar relationship of reaction rate to temperature of the reaction that a general description and equation describing the relationship has become widely used. The general subject is referred to as Arrhenius Reaction Rate Theory. The general equation for the Arrhenius Reaction Rate is written as:

$$k = Ae^{-B/T}$$

where **k** is the rate of reaction, **e** is the mathematical entity (approximately 2.718) that is the base for natural logarithms, **A** and **B** are constants that are determined by the reaction type, and T is the absolute temperature in Kelvins.

For the oil oxidation reaction, the reaction equation is such that the rate of reaction is changed by a factor of two every time the temperature changes by 10 Kelvins (a Kelvin represents the same temperature change as 1 °C). The situation can be summarized by the Figure 3.2, where the time for the oil to oxidize to a neutralization number of 0.25 mg KOH/g is compared at three temperatures – 70 °C, 60 °C, and (60 - X) °C (343 Kelvins, 333 Kelvins, and 333 - X Kelvins).

The time required to get to 0.25 acid number is twice as long for 60 °C as it is for 70 °C. Further, for the curve at 60 - X °C, the time is twice as long as it is for 60 °C. We refer to this relationship as the 10 °C rule – a ten degree decrease in temperature cuts the reaction rate in half, while a ten degree increase doubles the reaction rate. On the figure, X is therefore 10, and the temperature of the curve is 50 °C.

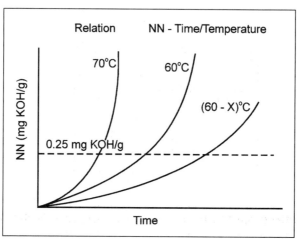

Figure 3.2 - Neutralization Number vs. Time

Other Accelerators – Moisture and Metals as Catalysts

To check out what effect some other possible "candidate accelerators" have on the oil oxidation, we need to devise an experiment. Let's adapt a procedure that is used to look at the oxidation stability of oil. The Doble Sludge Free Life (Figure 3.3) test is used to evaluate new oil for oxidation stability. In the Doble test procedure, the oil is artificially aged until the first signs of visible sludge are seen. It is based on a discontinued ASTM standard, D 1904, Standard Method of Test for Oxidation Characteristics of Mineral Transformer Oil.

In the first series of experiments, we are going to hold the temperature of the reaction at 95 °C while bubbling oxygen through the oil, so that the oil is pretty well saturated with the same dissolved oxygen content. We will run this for given time periods, and use the increase in neutralization number as a measure of how much the oil has oxidized. The time taken to complete the experiment and the neutralization number will be used to approximate a relative reaction rate. We are going to use two levels of moisture content (10 ppm and 50 ppm), and we are going to run the experiments with and without strips of copper and iron as catalysts.

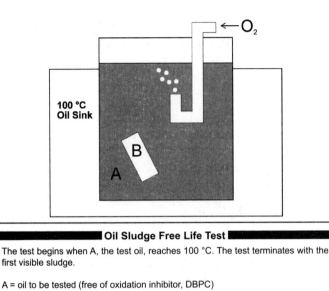

Figure 3.3 - A diagram illustrating the apparatus used to accelerate aging of the oil by bubbling oxygen through it at a controlled and elevated temperature.

Here are the results of our study of the effects of moisture and metal catalysts on the rate of oil oxidation: (see Table 3.1)

1. The first experiment is our control experiment – low moisture oil oxidized without a metal catalyst.
2. The second experiment changes the moisture content – other parameters remain the same. The oil oxidizes about 5.3 times as much in the same amount of time.
3. The third experiment adds an iron catalyst to our control experiment. The oil oxidizes about 3.8 times as much in the same amount of time.
4. If you were to predict what would happen if you added both high moisture and an iron catalyst to the control experiment, you might expect either 9.1 times the rate of oxidation (5.3 from Experiment 2 plus 3.8 from Experiment 3) or maybe even 20.1 as fast – 5.3 TIMES 3.8. What really happens is that you get 30 times the oxidation a little less

Catalytic Effects of Moisture and Transformer Metals

Experiment	Catalyst Metal	Water	Hours	Neut. Number
1. Control	None	Low	3500+	0.17
2. Moisture	None	High	3500+	0.90
3. Iron	Iron	Low	3500+	0.65
4. Iron and Moisture	Iron	High	400	8.10
5. Copper	Copper	Low	3000	0.89
6. Copper and Moisture	Copper	High	100	11.2

Table 3.1

than 12% of the reaction time. The reaction proceeds about 262.5 times as fast when both high moisture and an iron catalyst are present when compared to our control experiment.

5. The fifth experiment adds a copper catalyst to our control experiment. There is 5.2 times as much oxidation, but the reaction time has been reduced to 3000 hours. The reaction rate is about 6.1 times as fast due to the presence of the copper catalyst.
6. The final experiment has both high moisture and a copper catalyst. The same pattern is present as was there for iron – except more so. The oxidation rate when both high moisture and copper are present is about 2306 times as fast as the control experiment.

The rate of oxidation proceeds faster in the presence of higher water content. These types of experiments have been repeated and indicate that it is a direct relationship, as indicated by the 5:1 ratio of moisture content leading to approximately a 5:1 ratio of reaction rate. All other things being equal, if you double the water content, you double the rate of the oil oxidation reaction. (Another way to state this is that you cut the oxidation free life of the oil in half when the moisture is doubled.)

Iron and copper act as effective catalysts for the oxidation reaction. Copper accelerates the reaction more than iron does. Additional studies have indicated that the other metals usually found in transformers also act to catalyze oil oxidation. Bare solid metal and metal dissolved in the oil at high concentrations speed up the oxidation reaction.

The combined action of the metal catalyst and moisture is an example of synergy. The effects of two or more accelerants together are considerably more intense than would be expected of those catalysts apart. This will be extremely important to our maintenance and operations programs that are aimed toward life extension. Controlling those factors that we can control, because their action is combined with factors we cannot control, leads to what would seem to be a disproportionate improvement in the aging rate of the oil and the equipment. Or rather, it would seem to be disproportionate if we did not understand the concept of synergy and have this example concerning how it works.

Electrical Stress

The apparatus we used for the above series of experiments can be made to more closely mimic actual conditions in the transformer if the oil being aged is electrically stressed. Two Russian chemists – R. S. Lipshtein and M. I. Shakhnovich – reported the results of such an experiment in their 1970 book, *Transformer Oil*.

Figure 3.4 is a fair representation of the apparatus used. The two electrodes in the upper left portion of the diagram essentially form a small capacitor. The electrodes – or plates – are charged with a variable electrical potential measured in kilovolts.

At very low levels of electrical stress – the electrodes are charged with 10 kV/cm of separation – the following observations were noted:

- 20% more sludge deposits were formed in the same amount of time compared to a control experiment where the electrodes were not charged.
- Visible particles in the sludge deposits were much larger when the electrodes were charged compared to the control.

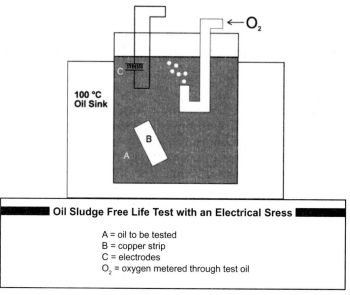

Figure 3.4 - A diagram illustrating the apparatus used to accelerate aging of the oil by bubbling oxygen through it with an electrical stress, at a controlled and elevated temperature.

- Deposits were much less uniform when the electrodes were charged compared to the control.
- Sludge deposits tended to accumulate in zones of electrical stress where the intensity was highest.

A second series was performed comparing a control with no electrical to moderate electrical stress caused by charging the electrodes with 49 kV/cm of separation. In this series, aging of the oil was performed for 44 hours and five parameters related to oil oxidation were measure for the control and for the higher stress environment. All five parameters indicated more extensive oxidation in the same period of time if the oil was stressed electrically. The following table (Table 3.2) summarizes these results.

1. Acid Number or Neutralization Number is a direct measure of the extent of oil oxidation.
2. Water Soluble Acid is a "subset" of the acid number. Water-soluble acids are stronger and lower molecule

Accelerated Aging of Oil Under Electrical Stress

Test Parameter	0 kV/cm stress	49 kV/cm stress	% increase comparison
Acid Number (mg KOH/g)	0.10	0.13	+30%
Water Soluble Acid (mg KOH/g)	0.032	0.049	+53%
Generated Moisture (weight %)	0.003	0.017	+466%
Tan Delta @ 70 °C	5.5	10.7	+95%
Oxygen Absorbed during Aging (mL O_2/100 g oil)	28.5	48.5	+70%

Table 3.2

weight but are also formed as a result of oil oxidation.

3. During the process of oxidation, water is generated from the reaction of oxygen with hydrogen released from the hydrocarbons during their degradation. Also, there is water chemically bound to the oil that is also released when the oil hydrocarbon chains oxidize.

4. Tan Delta is related to the liquid power factor and dissipation factor discussed later in this chapter. Like the acid number, tan delta is a relatively direct measure of oil oxidation.

5. The oil in the experiments is saturated with oxygen when the aging runs begin, and stays saturated throughout. As oxygen is used by the oxidation reaction, more oxygen is absorbed into the oil from the gas stream passing through it. If you measure the amount of oxygen used, that value is directly related to the amount of oil oxidation that has occurred.

So, the condition that is the whole reason that we have a transformer in the first place – to have a zone of controlled electrical stress that we can use to transform power to a more useful voltage – causes accelerated aging of the oil that we need to use in the insulating system.

Cellulose

The cellulose molecules that make up the paper in the solid insulation contribute to our oil aging properties.

- Paper acts as a catalyst. The oil oxidation reaction proceeds faster when paper is present than when it is not.
- Paper itself forms organic acids when it breaks down. Like the acids formed when oil oxidizes, these acids from the paper act aggressively to further tear apart the cellulose molecules.
- Paper is an **absorbent** material. Water and polar oxidation products "soak into" the paper. These compounds are attracted preferentially due to the chemistry of the cellulose molecules inside the structure of the paper.
- Paper is an **adsorbent** material. In addition soaking into the paper, moisture and oxidation products are preferentially attracted to coat the outside surfaces of paper, forming hydrogen bonds with the cellulose. These are weaker than "normal" chemical bonds, but are substantial enough to cause significant resistance to removing water and oxidation products from the paper surfaces.

Consider Figure 3.5 - an experiment in oil aging using no catalyst, a bare copper catalyst, and a copper catalyst wrapped in insulating paper.

There are two possible interpretations of these results. Either (1) The paper somehow "shields" the oil from the catalytic effects of the copper OR (2) The oil oxidation decay products are soaked up or otherwise removed by the paper so that they are not available in the oil to be detected by the Neutralization Number oil test.

When this experiment was investigated further, it was determined that the paper had absorbed and adsorbed oxidation decay products so that were not in the oil when the oil was tested. Other experiments, to more quantitatively look at the oxidation of oil in this type of aging, indicate that the paper contributes to more extensive and rapid oxidation of the oil.

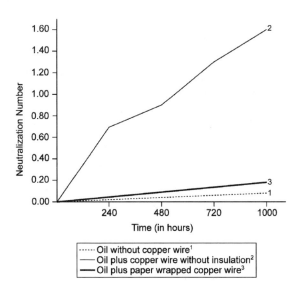

Figure 3.5 - Neutralization Number vs. Time with varying types of oil catalysts.

Oxidation Products

Once oil oxidation starts, the oxidation products themselves act as catalysts to speed up the oil oxidation reaction. There are a couple of instances where we can clearly see this.

Just about all oil oxidation experiments proceed along the same pattern. Oil oxidation proceeds gradually, at a relatively constant rate, until the oil oxidizes to the point where it has a neutralization of about 0.25 mg KOH/g. Beyond that point, which we refer to as the critical acid number, the rate of reaction increases markedly and continues to accelerate so that at higher acid numbers the rate of reaction gets much higher.

Also, if oxidation products are not completely removed during a maintenance procedure, the degradation and aging of the oil after the maintenance procedure is much faster than it had been originally when the oil and the insulating system were new. By contrast, the rate of aging is not any faster – and some experiments have indicated it may sometimes be slower – if the oxidation products are properly and thoroughly removed by maintenance.

Conditions in the paper are such that it is the preferred site for oxidation products to accumulate. For this reason, the structure of the paper itself is the place where aggressive oxidation products are first formed if proper maintenance to keep aging under control is not performed.

Controlling Accelerating Factors

A lot of the above discussion is not very good news. Conditions inside an operating transformer turn it into a huge and complicated chemical reactor. Just about everything that goes on inside that chemical reactor serves to speed up the reaction that we want to slow down – the reaction of oil with oxygen that causes the oil to age and to need maintenance.

We cannot do much about the presence of metal catalysts, cellulose in the paper, or electrical stress. We need to have these present in order for the equipment to do what it was designed to do. Maintenance can, however, control several of the other conditions that accelerate aging of the oil.

- Oxygen content can be reduced when the oil is installed, and ingress of oxygen during operation can be mitigated by the equipment design. Further, the dissolved oxygen content can be monitored during operation, and maintenance can be performed to reduce oxygen content if it increases to undesirable levels.
- Similarly, moisture content can be controlled during installation and operation of the equipment. Because moisture content of the solid insulation is even more destructive to the equipment than moisture content of the oil, it is even more imperative to regularly monitor the oil for moisture content and use those results to calculate moisture content of the solid insulation. Maintenance procedures are well established for reducing moisture in the oil and in the solid insulation.

- The operating temperature of the equipment can also be controlled. There are practical limits on the extent to which this can be done – you can only transfer so much load to other circuits. There are procedures such as keeping paint and bushings in good condition, properly ventilating installations, and controlling ambient temperatures, which have beneficial effects in terms of lowering operating temperatures. Properly controlling forced air or other cooling devices and even providing auxiliary cooling for equipment can be evaluated economically as temperature reduction strategies.
- Oxidation products can be removed – or, more accurately, not be allowed to build up in the oil and the solid insulation. Transformer oil testing is used to monitor the oxidation of the oil. Any changes in the results that indicate that oxidation products are present are the trigger to perform oil maintenance to clean the decay products out of the system.
- Oil life can be extended by the use of an inhibitor, where appropriate. Under most conditions, oil oxidation will not occur to any great extent as long as the oil continues to be inhibited. When inhibitor is depleted, it can be reintroduced using proper maintenance procedures.

Developing a Testing Program

When do you test? What do you test? What tests do you run?

Evaluate Your Testing and Maintenance Program

S. D. Myers personnel have been testing customer transformers since 1965. In our nearly 40 years, we have heard almost every "reason" possible for not testing electrical equipment insulating liquid. However, most of those reasons are not valid. More recently, a greater proportion of transformer owners are regularly testing the oil in their equipment. This change can be attributed to the better education of system maintenance managers resulting from the efforts of our Transformer Maintenance Institute and other similar efforts in the industry.

Why people test, though, is another story. Oil testing should be done as part of a well designed and implemented maintenance program. Oil test data that is obtained, but never evaluated or incorporated into a maintenance program, represents both money wasted and opportunity lost. Test data incorporated into a proper maintenance program improves operation and results in a more cost effective approach to meeting maintenance needs.

The way to answer the three questions presented at the top of this section comes as a result of critically evaluating the testing and maintenance program. Do test results determine when maintenance is done, and do they determine what maintenance is performed? If not, then the testing program does not serve any useful function except perhaps to give some warning before failure occurs. Only evaluation of the consequences and costs of failure or other effects of inadequate testing and maintenance provides enough information to determine whether a particular piece of equipment is worth testing and maintaining.

A system evaluation will define what kind of equipment is present and what the system maintenance managers' needs for information and data may be.

When do you test?

Establish a routine testing interval for testing each piece of equipment. *For most electrical equipment, that routine interval should be no longer than one year.* For some types of equipment where there can be more rapid changes in test data or where the costs of failure are particularly high, a shorter routine interval for some or all of the tests is recommended.

For example, furnace transformers almost always experience abnormally harsh operating conditions. Incipient fault conditions can develop much more rapidly in equipment in this type of application than would be expected for a normal distribution or power class unit of the same capacity. The oil testing program for furnace transformers should address this. In most cases, annual monitoring oil tests should be supplemented by quarterly testing for dissolved gases, at a minimum. In free breathing furnace transformers, additional quarterly or semi-annual testing for moisture contamination or oil oxidation may be indicated.

What do you test?

The true answer to this is that you need to test the equipment for which the results are going to be useful. If you actively maintain the equipment, testing the oil is imperative to tell you the proper time to perform maintenance. In cases where maintenance is never going to be performed, as the equipment is run until the end of its reliable life (which can be an appropriate management strategy in some cases), testing on a routine basis is performed to define when the equipment should be removed from service prior to failure.

If the equipment can be run to failure with minimal costs or consequences as a result of an unscheduled failure, the optimum choice may be not to test that piece of equipment. Utilities frequently make this reasoned management decision for distribution class equipment below a certain size.

To account for exceptions, great care is needed when making this type of decision. In our seminars, we often cite an example where an industrial facility was spending an unusually high portion of a transformer's replacement cost every year on annual and quarterly oil testing. When asked about the apparent waste of resources resulting from concentrating so much on an easily replaced unit, the responsible supervisor advised that this pad mounted transformer provided power to the owner's office and computer. The transformer in this service had failed once, with catastrophic results. The supervisor and the owner knew that the testing program was excessive, but they were willing to endure the expense to give them a better chance of preventing another unplanned outage for those systems that the unit powered.

What tests do you run?

The following are the recommended testing packages for various classes of power distribution equipment. When "Oil Screen" tests are referenced, the package includes the neutralization number, interfacial tension, D 877 dielectric breakdown voltage, relative density, color, and visual examination for appearance and sediment.

> *Oil-Filled Transformers – CriticalPac (Critical Transformer Package):*
> These include transformers with primary voltage class 230 kV and greater, more than 5000 gallons of dielectric liquid volume, applications such as generator step-up units, rectifiers and furnace transformers, TR sets, and any other transformer considered to be critical to the operation. Annual testing package includes Oil Screen, dissolved gas analysis, moisture, liquid power factor at 25 °C and 100 °C, inhibitor content, dissolved metals, and furanic compounds. Depending on the application or critical nature of the service that the unit provides, a quarterly dissolved gas analysis may be appropriate. For higher voltage class units and larger units, particularly any with a diaphragm conservator oil preservation system, the D 1816 dielectric is recommended. D 1816 may be a valuable option for other units if false "poor" results will not be an issue. Particle count distribution may be used to help diagnose whether poor D 1816 results actually represent a problem.

PowerPac (Power Transformer Package):
>These include substation power transformers and most transformers with primary voltage classes 69 kV and greater, but less than 230 kV. *PowerPac 1* – for initial in-service or baseline testing – includes Oil Screen, dissolved gas analysis, moisture, liquid power factor at 25 °C and 100 ° C, inhibitor content, and furanic compounds. *PowerPac 2* – for subsequent routine monitoring – includes Oil Screen, dissolved gas analysis, moisture, liquid power factor at 25 °C and 100 °C, and inhibitor content. Dissolved metals should be run routinely also if the unit has oil pumps. D 1816 may be a valuable option subject to the same concerns expressed above.

DistributionPac (Distribution Transformer Package):
>This package of tests is applied to distribution class equipment of any primary voltage class including pole mounted or pop-top transformers and pad mounted or cabinet transformers as well as most units with primary voltage class less than 69 kV. The annual testing package includes Oil Screen, dissolved gas analysis, and moisture. Other tests are run when routine monitoring indicates a need for further diagnosis of possible problems. Care should be taken when using this type of testing package. Smaller units where the test results are not used should not be tested if the resources can be more profitably used elsewhere in the testing program. Further, a critical transformer, even if of small size or primary voltage class below 69 kV, should be adequately tested to ensure proper maintenance.

Transformers Filled with Fluids Other than Oil – AskPac (Askarel Package):
>Annual test package includes Oil Screen, moisture, and furanic compounds. For askarel – as for most fluids other than transformer oil – the oil screen tests do not include the interfacial tension.

S FluidPac (Specialty Fluid – Non Mineral Oil Fluids):
> Examples – Silicone, natural and synthetic ester based fluids, and synthetic hydrocarbon fluids (other than high flash point hydrocarbon fluids that are tested the same as transformer oil). Annual test package includes Oil Screen (except for IFT), dissolved gas analysis, and moisture. Furanic compounds analysis is a valuable addition, especially to evaluate the apparent aging of the solid insulation or confirm a fault condition involving the paper. Liquid power factor is valuable in some cases, but should usually only be run at 25 °C for these fluids.

WecPac (Perchloroethylene fluids including OEM fluids such as Wecosol and Askarel retrofill fluids):
> Annual test package includes Oil Screen and moisture. Furanic compounds analysis is an option as discussed immediately above. For units that were initially askarel and that were retrofilled with perchloroethylene which remains as the permanent dielectric fluid, the fluid should also be tested annual for the acid scavenger AGE (allyl glycidyl ether). A retrofilled askarel unit that still contains perchloroethylene where the acid scavenger has been depleted should be replaced.

Other Equipment – LTCPac (Oil Filled Load Tap Changers):
> Annual test package includes full Oil Screen, dissolved gas analysis, moisture, and particles and filming compounds analysis. Units with abnormally high rates of operation may require more frequent dissolved gas and particle analyses. LTCs with on-line filters do not require Particles and Filming Compounds analysis.

RegPac (Oil Filled Regulators):
> *RegPac - Single:* Annual test package for single-phase regulators includes Oil Screen, dissolved gas analysis, and moisture.

RegPac - Three: Annual test package for three phase regulators includes Oil Screen, dissolved gas analysis, moisture, liquid power factor at 25 °C and 100 °C, inhibitor, dissolved metals, and furanic compounds analysis.

RegPac - Step: Annual test package for step voltage regulators includes Oil Screen, dissolved gas analysis, moisture, liquid power factor at 25 °C and 100 °C, inhibitor, dissolved metals, furanic compounds analysis, and particles and filming compounds analysis.

OCBPac (Oil Filled Circuit Breakers):
Annual test package includes Oil Screen, dissolved gas analysis, moisture, and particle count distribution.

SwitchPac (Oil Filled Switches):
Annual test package includes Oil Screen, dissolved gas analysis, and moisture. Particle count distribution is recommended. <u>NOTE:</u> Before testing oil filled switches, be advised that some manufacturers have recommended removal and total replacement of their oil filled switches due to the hazard potential of explosions when the switch fails.

For any equipment or apparatus not listed, a minimum annual recommendation for testing includes Oil Screen, dissolved gas analysis, and moisture.

Testing In-Service Transformer Oil
Routine Monitoring Tests

Liquid Power Factor (Dissipation Factor)
ASTM Standard Method D 924

Liquid power factor is an excellent test for monitoring in-service transformer oil. The test is valuable for assessing new oil from a supplier and for evaluating new oil installed in equipment. As the oil continues in service, there are a number of conditions that degrade the oil that show up as changes in the liquid power factor results.

When a dielectric liquid such as transformer oil is subjected to an alternating current field, there are dielectric losses which cause two effects. The resulting current is deflected slightly out of phase with the AC field that has been applied, and the energy of the losses is dissipated as heat. Liquid power factor and the closely related dissipation factor are direct measures of these dielectric losses. (Liquid power factor is calculated as the sine of the loss angle – the amount of current deflection due to dielectric loss – while the dissipation factor is the tangent of the same loss angle.)

New, clean, and dry transformer oil has a very small value for liquid power factor. Contamination of the oil by moisture or by many other contaminants will increase the liquid power factor. The aging and oxidation of the oil also elevates the liquid power factor values. Therefore, this is an extremely useful test because almost everything "bad" that can happen to the insulating oil will cause the liquid power factor to increase.

When liquid power factor determinations are run on transformer oil, they are usually run at two temperatures: 25 °C and 100 °C. This is done because the two readings, and how they change over time, can be useful in diagnosing which condition (moisture, oil oxidation, or contamination) is causing a high power factor. Further, the 100 °C value is many times more sensitive to small changes in oil characteristics.

Liquid power factor values are usually small numbers – so much so that the convention in the United States is to report the liquid power factor as a percentage. As an example, consider new oil installed in a new transformer of primary voltage class less than 230 kV. The 25 °C liquid power of that oil should be no more than 0.0005 (0.05%) which is the recommended test limit for this value in ANSI/IEEE C57.106-2002. Frequently, in actual new installations, the measured liquid power factor at 25 °C is much lower. Reporting the percentage is considered less cumbersome.

In-service values for classifying liquid power factors are the same for all primary voltage classes of equipment. For oil filled transformers, those values are:

Liquid Power Factor Values

	AC	QU	UN
@ 25 °C	< 0.1%	≥ 0.1% ≤ 0.3%	> 0.3%
@100 °C	< 3.0%	≥ 3.0% ≤ 4.0%	> 4.0%

Dissipation factor is not exactly the same value as liquid power factor, but the two measures are close enough throughout the ranges of values that we see with transformers so that we can essentially use them interchangeably. For example, if the liquid power factor is 4.0%, the dissipation factor is 4.003%. As the values get larger, the difference also gets larger. Even at a huge liquid power factor of 10.0%, though, the difference is still very small (the dissipation factor is 10.05%).

The 100 °C value is always considered more carefully than the 25 °C value when determining whether there is a problem. QU or UN values for liquid power factor should be investigated and the cause should be diagnosed. Reclaiming the oil or hot oil cleaning the transformer will reduce the liquid power factor. Drying out the oil will also improve the liquid power factor, particularly the 25 °C reading.

Figure 3.6 - Liquid Power Factor test set

A "first time" bad reading may not indicate a problem, though. Care should be taken before committing to maintenance based on a high liquid power factor reading. Chemical changes in the oil may occasionally cause elevated liquid power factor values. These high readings are usually temporary and are not of concern. If elevated values persist – the rule of thumb we use is if they continue to appear in the next annual testing – then the cause should be identified and corrected, if appropriate. Extremely elevated liquid power factors at 100 °C in transformer oil usually indicate contamination. Care must be taken when dealing with fluids other than transformer oil because some of these either have naturally high liquid power factor values when new or develop such high values very soon after being put in-service. For these fluids, liquid power factor is not a very good test for monitoring in-service degradation of the oil.

The way the test is run is that oil is placed in a test cell. The test cell consists of an inner and outer shell with a small gap that the oil fills up. When the two cells are energized by an AC current, the thin film of oil is subjected to the AC field resulting in the dielectric losses that the instrument can measure and report as either liquid power factor or dissipation factor.

The same apparatus can be used on completely clean and brand new insulating liquid to measure the dielectric constant. The dielectric constant is used as a design parameter as it is an indication of how well the insulating liquid will act as a dielectric by itself to insulate conductors from each other, when used in series with a solid dielectric, or when used as the insulation medium in a capacitor. Also, there is a similar test – resistivity, which is performed using ASTM D 1169 – that measures the specific resistance of the insulating liquid to a DC electric potential.

Moisture in Oil
ASTM Standard Method D 1533

This test method determines the moisture content of insulating oil using an automatic coulometric Karl Fischer titrator. A specimen of the oil is injected into this device, and the device adds reagents automatically until the endpoint is reached. The endpoint is determined by electrodes that sense electrical conditions in the reaction flask. When the endpoint is reached, the device stops the titration and electronically calculates the moisture content of the oil from the volume of oil injected and the amount of reagent consumed. Moisture content of the oil is reported in parts per million (milligrams of moisture per kilogram of insulating liquid).

By itself, particularly for mineral oil filled transformers, the moisture content in parts per million (ppm) is not sufficient to evaluate moisture content of in-service oil. The ppm value is useful in evaluating new oil from a supplier or for installing new, processed oil in equipment. The moisture content in ppm is also of primary importance for in-service fluids other than oil, and may be used as a criterion for mineral oil filled equipment other than transformers. Most of the time, though, for oil filled equipment and especially for mineral oil filled transformers, the moisture in parts per million is only a small part of the information that needs to be considered.

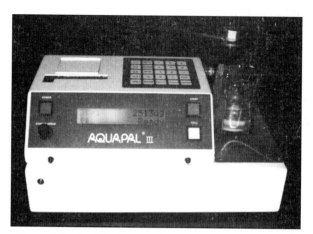

Figure 3.7 - Karl Fischer test set.

Moisture in electrical equipment presents two conditions that are detrimental. First, moisture raises the risk of dielectric failure in the equipment. The most serious condition that could occur would be if there were sufficient moisture present to cause free water to come in contact with energized conductors, as this would lead to immediate and catastrophic failure. Second, moisture contributes to accelerated aging of the insulating liquid and of the solid insulation. Moisture degradation of the solid insulation causes permanent damage and premature loss of equipment life.

These two effects are addressed by how we use the moisture values when evaluating moisture content of oil filled transformers. The risk of dielectric failure is related directly to the percent relative saturation of the oil with water. The permanent premature aging of the solid insulation is directly related to the percent moisture content of the solid insulation. The moisture content values that are obtained from the Karl Fischer titration are used to calculate percent saturation in the oil and percent moisture by dry weight in the solid insulation.

Percent Saturation

Moisture is not very soluble in new, clean transformer oil. The solubility of water in oil is higher at higher temperatures. Comparing how much moisture is dissolved in the oil to how much moisture the oil can hold is what we refer to as the relative saturation of the oil. For example, oil that is 40 °C will hold a little more than 120 ppm of moisture in solution. If the actual moisture content is 12 ppm, the relative saturation is 10%. The percent saturation of the oil is calculated from the ppm moisture content and the oil temperature.

If the moisture in the oil is higher than the desired relative saturation and the transformer cools significantly, some of the dissolved water can come out of solution as droplets of free water. Those droplets represent a condition that can cause immediate dielectric failure if they come in contact with an energized conductor inside the equipment. IEEE Standard C57.106 – 2002 includes limits on maximum saturation for oil filled equipment, and those limits are based on voltage classes. For equipment with primary voltage class up to and including 69 kV, the IEEE limit for continued use is a maximum saturation of 15%. For primary voltage class between 69 kV and 230 kV, the limit is 8% saturation. For primary voltage class of 230 kV or greater, the maximum limit is 5% saturation. According to the IEEE standards, it is beyond these levels that the oil presents an unacceptable risk of dielectric failure due to moisture in the oil.

We use these IEEE limits as the maximum acceptable values in our classification system. Above these limits are questionable ranges for each voltage class, indicating values that should be addressed by appropriate maintenance procedures. If moisture in the oil is not addressed appropriately, the values may increase beyond these questionable ranges and become unacceptable. At these levels, immediate maintenance is required as the risk of immediate dielectric failure is high and potentially catastrophic.

The ranges used for percent saturation by primary voltage class are as follows:

Percent Saturation Values

Voltage Class	AC	QU	UN
≤ 69 kV	≤ 15%	> 15% < 20%	≥ 20%
> 69 kV < 230kV	≤ 8%	> 8% < 12%	≥ 12%
≥ 230 kV	≤ 5%	> 5% < 7%	≥ 7%

Percent Moisture by Dry Weight

Moisture in the paper that makes up the solid insulation is primarily a concern because it ages the insulation prematurely, leading to a shortened effective lifetime for the equipment. In rarer cases, there is high enough moisture content in the solid insulation to cause the insulation to degrade in the form of shedding fibers. If moisture is allowed to exceed even these levels, the transformer can get bad enough that there may be electrical flashover at temperatures frequently encountered in operating transformers. Moisture by dry weight values are also calculated from the ppm moisture content of the oil and the oil temperature. Rather than just setting acceptable, questionable, and unacceptable limits, there is a more useful grading system from "A" to "F" for percent moisture by dry weight values.

The top end or "most wet" limit of the "A" range represents the maximum percent moisture by dry weight content of the insulation paper where accelerated aging has not yet begun. At all levels above this range, moisture should be removed by suitable field dry out procedures. This task becomes progressively harder and more expensive as the moisture levels go up. The "D" range represents a practical limit for removing moisture in a cost effective manner – if moisture advances into the "F" range, the only practical response is equipment replacement.

As is the case with percent saturation, evaluating percent moisture by dry weight values has to take the voltage class of the equipment into account. IEEE Standard C57.106-2002 has limits for continued use based on percent moisture by dry weight. For equipment with primary voltage class up to and including 69 kV, the IEEE limit for continued use is a maximum of 3% moisture by dry weight. For primary voltage class greater in between 69 kV and 230 kV, the limit is 2% moisture by dry weight. For primary voltage class of 230 kV or greater, the maximum limit is 1.25% moisture by dry weight. As is evident from the table below, these limits, falling as they do into the "D" range of grades, represent the upper limit where moisture may still be practically removed. These levels are well into the area where damage to the solid insulation by premature aging is occurring.

The grading system for percent moisture by dry weight values in solid insulation is as follows:

Percent Moisture by Dry Weight Values

Voltage Class	A	B	C	D	E
≤ 69 kV	0 to 1.25%	1.26% to 2.00%	2.01% to 2.50%	2.51% to 4.00%	4.01% and up
> 69 kV < 230kV	0 to 0.85%	0.86% to 1.35%	1.36% to 1.70%	1.71% to 2.65%	2.66% and up
≥ 230 kV	0 to 0.55%	0.56% to 0.85%	0.86 to 1.05%	1.06% to 1.70%	1.71% and up

Moisture values, especially the calculated values of percent saturation and percent moisture by dry weight, are very subject to fluctuations based on sampling conditions, ambient temperature conditions, and fluctuations in transformer load and operating temperature. A first time determination that a transformer is wet may be due to one of these conditions. If moisture results suddenly indicate a wet transformer – for example, a first test or one that is inconsistent with past history – it is frequently more responsible to retest the transformer at a reduced interval of no more than three months before concluding that a dry out procedure is required.

Oxidation Inhibitor
ASTM Standard Method D 2668 or D 4768

2,6-ditertiary-butyl para-cresol (DBPC) and 2,6-ditertiary-butyl phenol (DBP) are used as oxidation inhibitors in transformer oil. Use of an oxidation inhibitor in the oil is recommended for equipment without adequate oil preservation systems where the dissolved oxygen content exceeds 1000 ppm. Testing for the oxidation inhibitor content of in-service oil is very important. Frequently, depleted inhibitor is the first indication that oil maintenance is needed. Under most conditions, the oil will not begin to age substantially due to oxidation if there is sufficient oxidation inhibitor present.

There are two standard test methods for oxidation inhibitor. Both methods detect the two compounds, DBPC and DBP, which are used as antioxidants in transformer oil and report their combined content as total oxidation inhibitor content. The D 2668 method uses an infrared spectrophotometer to determine inhibitor content, while the D 4768 method uses gas chromatography. Both methods yield equivalent results. Choice of method used depends on availability of instrument time in the laboratory. Oxidation inhibitor content is reported as weight percent of total inhibitor in the oil.

Figure 3.8 - Equipment used in measuring inhibitor content.

The classification of results for oxidation inhibitor content (weight percent) is as follows:

Oxidation Inhibitor Content

AC	QU	UN
≥ 0.2%	≥ 0.1% < 0.2%	< 0.1%

The optimum level for oxidation inhibitor is 0.3% in oil where use of inhibitor is recommended. Inhibiting the oil is always recommended, regardless of expected dissolved oxygen content, if the oil has been maintained using reclamation procedures.

Liquid Screen Tests - Oil Screen Package

In 1957, at an ASTM symposium on electrical insulating oil, a battery of seven tests was presented as the recommended in-service oil tests for electrical equipment. The recommended test package included the liquid power factor (run at 25 °C, only) and six tests that continue to comprise what is generally referred to an "oil screen package" or as "liquid screen tests." As is evident from this section of the *Guide*, we have learned a lot more about testing transformer oil since 1957. While the six oil screen tests continue to provide valuable information, they are not considered to be completely adequate for monitoring in-service oil.

The six oil screen package tests include:
- Neutralization Number (Acid Number)
- Interfacial Tension (IFT)
- Relative Density (Specific Gravity)
- Color
- Appearance (Sediment)
- D 877 Dielectric Breakdown Voltage

Neutralization Number

ASTM Standard Method D 974, D 664, or D 1534

When oil oxidizes and ages in service, some of the oxidation decay products that are formed are acidic in nature, meaning that they will react with and be neutralized by a base (alkaline) material. The neutralization number (acid number) has been used to evaluate transformer oil and other petroleum products since at least the early part of the 20th century. (The standard methods for acid number also include a determination of base number for petroleum products, but this is not used as an in-service test for transformer oil.) The standard methods for acid number use potassium hydroxide (KOH) to react with the acidic compounds in the oil. The amount of KOH needed to react with all of the acidic compounds is noted by either a color change of an indicator included in the oil/reagent mixture or by an electrical change measured by electrodes. Acid number is reported as milligrams KOH per gram of sample.

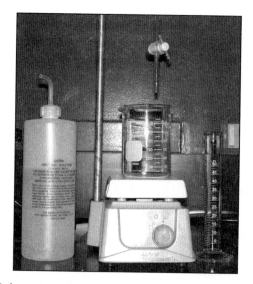

Figure 3.9 - Laboratory equipment (50 ml graduate cylinder, burette and beaker) and titration solutions are required to perform ASTM D 974 neutralization (acid) number test.

Classification of Neutralization Number results is as follows:

Neutralization Number Values

	AC	QU	UN
mg KOH/g sample	≤ 0.05	> 0.05 ≤ 0.10	> 0.10

Questionable or unacceptable results indicate that the oil has oxidized. A first test in the QU or UN range, or a sudden change exceeding the acceptable limit of 0.05 mg KOH/g, should be confirmed by a retest. Otherwise, if the acid number value is the result of a visible trend in the results over time or if a QU or UN value is confirmed by retest, the oil and the equipment should be hot oil cleaned.

Standard Method D 974 is the most often-used laboratory test, particularly the manual titration. The endpoint of the titration is defined by a color change for an acid/base indicator that is included in the reaction vessel. The indicator most often used is colorless when there is no excess base material in the reaction vessel and turns pink in the presence of excess potassium hydroxide. The titration continues adding small quantities of KOH until the mixture of oil and indicator turns pink. Then, a simple calculation yields the acid number from the amount of the potassium hydroxide reagent used and the weight of the test specimen. Many laboratories favor the manual titration method of D 974 because it is relatively simple and precise. The laboratory equipment and reagents required are also relatively inexpensive.

The D 974 has also been automated, using an automatic titrator to add the potassium hydroxide until the endpoint is reached. In the most widely used version of this automated test, pH electrodes are used to define the endpoint. A problem that a number of labs experience with this test is that the instrumentation is calibrated using too low of a pH to define the endpoint. The result of improper calibration of the instrument is that the acid numbers reported are unrealistically low, frequently staying below 0.005 mg KOH/g for in-service oil. Normally, we would expect to see values in the vicinity of 0.01 or 0.015 mg KOH/g for new installations and 0.01 to 0.04 or 0.05 mg KOH/g for properly maintained in-service oils.

To address this problem while maintaining the benefits of using automated equipment, some laboratories are using a colorimetric automatic titrator where a photosensitive cell is used to identify the color change in an automated color indicator titration. Such a standard method has been introduced to both ASTM and to IEC as a possible addition to the standard method for acid number determination.

Standard Method D 1534 for approximate acidity is very similar in concept to the D 974 manual titration. Instead of making a drop-by-drop addition of potassium hydroxide to get a precise value for neutralization number, increments of potassium hydroxide are used so that the "approximate acidity" can be inferred as being between two of the incremental additions. This approximate acidity test works well for field evaluations of oil aging and oxidation. Test kits have been manufactured to make use of the test in the field even easier.

Standard Method D 664 is another automated test for acid number. The method is similar in that potassium hydroxide is added to the reaction vessel by an automatic titrator, but the endpoint is defined differently. During the titration, potentiometric electrodes monitor the electrical environment in the vessel. The response of these electrodes in terms of the electrical potential of the reaction is graphed during the addition of the potassium hydroxide. At some point, the electrical potential undergoes a sudden change from a gradual slope to a more pronounced slope. This is referred to as the inflection point, and the amount of potassium hydroxide used to get to this inflection point is used to calculate the acid number. In the past, the D 664 method has been difficult to calibrate and to use. Changes in the method and improvements in the equipment available have broadened the use of this method considerably.

Results from either of the quantitative methods for acid number, D 974 or D 664, should generally be equivalent.

Interfacial Tension (IFT)
ASTM Standard Method D 971 or D 2285

Materials that do not mix readily form a surface or interface when they are brought into contact. This happens so frequently that we do not usually even think about it. Examples are the fact that water poured into a container forms a surface where it meets the air and soap bubbles form a surface between the soap solution and the air inside them (as well as outside). The surface that forms between air and water is strong enough to support a steel needle if it is carefully floated on the surface – this is a frequently used (and very impressive) junior high science demonstration of a physical principle.

Oil and water will not mix, either. If you put transformer oil and water in contact with each other, they separate and form an interface between the two dissimilar liquids. Like the surface of water in contact with air, the interface between oil and water is a real barrier. It takes some force to move an object from one phase to

Figure 3.10 - Interfacial tension is one of the tests used to monitor oil aging. Since oxidation products have an affinity for water molecules, this test is a logical one. ASTM D 971 laboratory test uses the Du Nouy tensiometer, a platinum ring, a glass container, and distilled water. Once the instrument is set up and calibrated, the test can be conducted in a few minutes. The term "tensiometer" refers to the instrument devised by Dr. Pierre Du Nouy. The ring method was known before the U.S. Civil war. Du Nouy's contribution was the development of a specialized form of torsion balance which greatly simplified the test and the calculations.

another across the interface. The strength of the barrier between air and water is referred to as surface tension. For the barrier between oil and water, the term becomes interfacial tension, usually abbreviated as IFT.

The practical use of IFT can be described very simply. Clean, new, and well-refined transformer oil has a relatively high IFT, measured and reported as millinewtons per meter (or as dynes per centimeter, an essentially equivalent unit). As the oil ages and oxidizes, the polar compounds that are formed by oxidation weaken the interface, reducing the interfacial tension. This occurs because the polar molecules are partially soluble in both the oil and in the water, so individual molecules are oriented along the interface.

The operation of the device used to measure the interfacial tension is cross-checked against the surface tension between air and distilled water. That surface tension is in the range of 71 to 73 mN/m. New, highly refined oil is generally in the range of 45 to 50 mN/m for IFT. When installed in the equipment, the new oil IFT will be reduced by about 5 to 10 mN/m as the oil picks up contamination from the inside of the new equipment. IFT will then gradually trend downward as the oil ages.

There are two standard methods that give essentially equivalent results. The ring method, ASTM Standard Method D 971, measures the IFT using a tensiometer such as that devised and introduced by Dr. Pierre Du Nouy. The water drop method, ASTM D 2285, calculates the IFT from the size of water drops that will remain suspended from a needle immersed in the oil being measured.

The classification for interfacial tension results is as follows:

Interfacial Tension Values

	AC	QU	UN
mN/m (dynes/cm)	≥ 32	< 32 ≥ 28	< 28

Similar to acid number, IFT values are used to define needs for oil maintenance. A "first time" or inconsistent low IFT should be confirmed. If the result of a continued trend downward, or if a low value is confirmed, a QU or UN IFT result should be addressed by hot oil cleaning of the oil and the equipment.

Ring Method D 971

The basic procedure of the method involves measuring the amount of force it takes to move a platinum ring of known dimensions from the water phase into the oil phase. A precision tensiometer is needed to accurately measure that amount of force and convert it to an equivalent tension measurement based on the dimensions of the ring.

The most common procedure using a tensiometer involves checking the calibration of the device against a clean sample of distilled water and then floating a test specimen of the oil on top of the oil. This uses a beaker that sits on a movable platform. The ring from the tensiometer is inserted into the beaker, below the

Figure 3.11 - A transformer oil lab technician carefully conducting an interfacial test on an oil drawn from a transformer only hours earlier.

interface and into the water phase. Slowly, the platform is lowered which has the effect of moving the ring from the water phase into the oil phase. The slow movement of the ring distorts the interface, essentially stretching the water up into the oil. Eventually, the pull on the interface exceeds the strength of the barrier, and the interface ruptures. The value indicated on the tensiometer dial at the time of rupture is either a direct reading of the interfacial tension, or it can be converted to the interfacial tension by a simple calculation.

D 971 is a "non-equilibrium" method. The oil/water interface is formed and allowed to age for 30 seconds before the movement of the ring is begun and the test must be completed so that approximately 30 more seconds are used to complete the movement of the ring and rupturing of the interface. The entire test is completed within the range of 50 to 70 seconds. There are automatic tensiometers being marketed that are intended to be equivalent to this non-equilibrium method. At the time of this writing, we are not familiar enough with them to express an opinion. In the past, automated devices for an "equilibrium" method, where the interface was aged sufficiently long as to come to more of an equilibrium state, have been introduced. To date, none of these have been demonstrated to meet equivalency with the standard ASTM methods.

Water Drop Method D 2285

The equivalent water drop method forces distilled water from a precision syringe through a needle that has been inserted into an oil test specimen. The size of the drop of water that will remain suspended from the needle without breaking off and sinking to the bottom of the container is related directly to the interfacial tension of the oil. To provide equivalent "non-equilibrium" conditions, an approximation is first run and then a drop using about three-quarters of the expected volume is allowed to age for 30 seconds. The remainder of the drop is then expressed out of the syringe so that the drop breaks off of the needle within 40 to 60 seconds of the start of the test. A simple calculation is used to convert the volume of water to interfacial tension. The values are equivalent to those obtained by the ring method.

Relative Density (Specific Gravity)
ASTM Standard Method D 1298

Relative density is more commonly (but slightly less accurately) referred to as "specific gravity." Relative density is a physical property of the insulating oil and is simply the ratio of the mass of a specific volume of the oil to the mass of the same volume of water at the same temperature. The test is very simple to run. Oil is placed in a cylinder and a device called a "hydrometer" is floated in it. The hydrometer is weighted so that it partially sinks in the oil being tested. The surface of the oil crosses the hydrometer at one of the marked divisions on the hydrometer's scale (actually, the defining point is the bottom of a curved portion of the surface referred to as the meniscus). That marked division indicates the relative density directly.

Naphthenic transformer oil has a relative density between 0.84 and 0.91. Most oils that are actually in service fall into a narrower range of about 0.86 to 0.89. Values lower than 0.84 typically indicate that oil is a paraffinic oil, and some synthetic oils, particularly the synthetic isoparaffinic oils, also fall in this range. Values over 0.91 indicate contamination by higher density materials, and a normal "suspect" for such contamination is PCBs.

The classification for relative density is as follows:

Relative Density

AC	QU	UN
0.84 to 0.91	< 0.84	> 0.91

Relative density is a calculated ratio and has no specific units of measurement.

Relative density usually does not change as the oil is in-service because aging and oxidation have little effect. Values outside of the acceptable range or significant changes between normal monitoring intervals are of concern, and the cause should be identified.

Contamination by oils of different relative density may be noted, but the changes are often too subtle to be completely evident from the previous relative density values if the relative density of the contaminating material is close to that of transformer oil. If such contamination is suspected, confirmation can come from other tests such as flash point determination, viscosity measurements, infrared spectrophotometry, or other more quantitative methods of analysis.

Color
ASTM Standard Method D 1500 or D 1524

ASTM Standard Method D 1500 is a laboratory determination of the color of petroleum products. ASTM Standard Method D 1524 is a method for visual examination of electrical insulating oils in the field that includes estimation of the ASTM color. Both methods involve comparing an oil sample to tinted glass standards and reporting the closest match as the ASTM Color on a scale of 0.5 to 8.0. The units are arbitrary ASTM Color units. Results from the two methods are very similar, but not identical. Typically, the results are close enough to being equivalent that the difference does not affect any management decisions concerning the oil and the equipment.

The color of new transformer oil is very low. Visually, clean, new oil is almost water white and completely clear. (The color of the oil in such a case is reported as "less than 0.5.") As the oil ages and oxidizes, it gets darker. Contamination may also cause a rapid change in color.

Taken by itself, color has little meaning. "Bad" oil can be lightly colored, while dark oil can still be of like new quality in all other respects. Poor color of the oil rarely affects the performance of the oil in service. Further, reclaimed oil that meets or exceeds all other performance and quality criteria may still be somewhat dark in color. In general, very dark oil is investigated for cause, as are large changes in color from one routine sampling period to the next.

Classification of color results is as follows:

ASTM Color

AC	QU	UN
≤ 3.5	--	> 3.5

Visual Examination of Insulating Oil
ASTM Standard Method D 1524

The oil sample is visually checked for cloudiness, turbidity, suspended particles, visible sediment or sludge, carbon, free water, or anything that is not clear, homogeneous transformer oil. Acceptable oil will be clear and bright, free from any visible contamination or presence of any of the previously listed abnormal conditions.

Anything other than "clear and bright" is unacceptable and should be investigated to determine the cause. Cloudiness or turbidity indicates suspended water droplets, carbon, or sludge. Visible carbon indicates that there probably has been arcing in the equipment. This condition should be investigated further using diagnostic tests such as dissolved gas analysis or ICP analysis for dissolved metals. Sediment or any visible particles can be analyzed microscopically to determine their source.

Dielectric Breakdown Voltage
ASTM Standard Method D 877

There are two ASTM standards for dielectric breakdown voltage of insulating oil, D 877 and D 1816. The D 877 method measures the breakdown voltage using a test cell that has two flat disk electrodes spaced 0.10 inches apart. A dielectric breakdown voltage test is run by subjecting the two electrodes to a steadily increasing electrical potential until there is a discharge through the liquid being tested from one electrode to the other. For the D 877 test, the voltage over the electrodes is increased by 3,000 every second until the dielectric breakdown occurs. Typical D 877 readings for in-service insulating oil are from 30 to about 60 kilovolts.

The purpose of performing dielectric breakdown voltage determinations is to evaluate the oil's ability to withstand electrical stress. Contamination of the oil due to such things as fibers from the solid insulation, conductive particles, contamination by foreign matter, dirt, and water can affect dielectric breakdown voltage. The utility of the D 877 test is limited because the test is not sensitive to moisture unless the moisture content exceeds 60% of the saturation level. Further, the D 877 test is not sensitive to oxidation products in service aged insulating oil.

The D 877 dielectric breakdown voltage method was widely accepted in the past, and there was great reluctance to admit that the usefulness of the test was limited. Over the course of the last few years, acknowledgements of the method's limitations have become more widely accepted. When IEEE standard C57.106 – 2002 was published, all references to D 877 as an in-service oil test for mineral oil in transformers were eliminated in favor of limits for the D1816 dielectric tests. Many laboratories and owners have eliminated use of the D 877 test, entirely.

Figure 3.12 - TC/VDE: ASTM test cell with VDE electrodes and motor driven circulating system.

The value from a D 877 determination still has meaning, however, and is a valid tool as long as the limitations and relative insensitivity of the test are acknowledged. The D 877 method will still indicate the presence of some types of contamination in mineral oil filled transformers. Also, for equipment other than transformers that typically exhibit higher moisture content or where there may be metal particles present and for equipment filled with fluids other than mineral oil, the D 877 test still provides useful information.

The D 877 dielectric breakdown voltage is still usually included in oil screen test packages.

Classification of dielectric breakdown voltage results using the D 877 method is as follows:

Dielectric Breakdown Voltage
Disk Electrodes D 877 Method

	AC	QU	UN
kilovolts (kV)	≥ 30 kV	< 30 kV ≥ 25 kV	< 25 kV

Dielectric Breakdown Voltage
ASTM Standard Method D 1816

This standard method of measuring dielectric breakdown voltage uses spherical VDE (Verband Deutscher Elektrotechniker) electrodes. The method is run at one of two gap settings, 1 millimeter (0.04 inches) or 2 millimeters (0.08 inches). The D 1816 method for determining dielectric breakdown voltage is more sensitive to moisture and is also sensitive to polar compounds such as oil oxidation products. Sensitivity to some particles, particularly insulation fibers, is much more consistent with this method.

Because of the greater sensitivity, the rate of voltage rise is lower – 500 volts per second. Also, the D 1816 test cell has a motor driven agitator. This agitator runs during the test to cause the oil to flow between the electrodes, carrying suspended particles into the gap between the VDE spheres where they can affect the breakdown voltage.

Originally, the method was intended strictly for equipment of primary voltage class 230 kV and above where the oil had been processed with vacuum and filtration before being installed. Over many years, a number of labs started using the method to evaluate in-service oil from equipment of lower voltage classes. In 2002, the revision of IEEE Standard C57.106 incorporated D 1816 limits for new and in-service oil.

The classification for D 1816 dielectric breakdown voltage results depends on the primary voltage class of the equipment and is as follows, for the two gap settings (dielectric breakdown voltages are in kilovolts):

D 1816 Dielectric Breakdown Voltage
1 mm gap setting

Equipment Voltage Class	AC	QU	UN
≤ 69 kV	≥ 23 kV	< 23 kV ≥ 18 kV	< 18 kV
> 69 kV < 230kV	≥ 28 kV	< 28 kV ≥ 23 kV	< 23 kV
≥ 230 kV	≥ 30 kV	< 30 kV ≥ 25 kV	< 25 kV

D 1816 Dielectric Breakdown Voltage
2 mm gap setting

Equipment Voltage Class	AC	QU	UN
≤ 69 kV	≥ 40 kV	< 40 kV ≥ 35 kV	< 35 kV
> 69 kV < 230kV	≥ 47 kV	< 47 kV ≥ 42 kV	< 42 kV
≥ 230 kV	≥ 50 kV	< 50 kV ≥ 45 kV	< 45 kV

In addition to being sensitive to moisture, particles, and contamination, the D 1816 dielectric breakdown voltage method is also sensitive to the presence of dissolved gases. This limits the usefulness of this method also, although it is more widely applicable than the D 877 method. A "good" D 1816 result indicates that there is nothing seriously wrong with the dielectric breakdown strength of the oil. A "bad" D 1816 result, however, does not always indicate that there is something wrong with the oil. Acceptable levels of the other materials to which the test is sensitive can be detected along with a normal amount of dissolved gas to depress the D 1816 result so that it is no longer in the acceptable range.

Depending on the D 1816 to provide meaningful results for smaller industrial and distribution class equipment that is gas blanketed is usually not terribly useful. Frequently, the D 1816 results indicate there may be problems with the dielectric breakdown voltage when those problems are not real. Even for higher voltage classes, or larger equipment where dissolved gas levels are much reduced, where the D 1816 may be recommended as a monitoring test, care must be taken when interpreting D 1816 results to ensure that resources are not expended to chase down problems that do not exist except as a peculiarity of the test method.

Monitoring and Diagnostic Tests for Transformer Oil

Dissolved Gas Analysis
ASTM Standard Method D 3612

Dissolved gas analysis is a critical test for monitoring the mechanical and electrical operation of electrical equipment. It is particularly useful for diagnosing fault conditions and operational problems with transformers and load tap changers.

Testing insulating oil for dissolved gases and interpreting the results of that dissolved gas analysis are such critically important tools for maintenance of transformers that this topic is covered by its own section in this *Guide*, starting on page 113.

Dissolved Metals Analysis
ASTM Standard Method D 3635 and Pending ASTM Method for ICP Analysis of Metals in Oil

ASTM Standard Method D 3635 is for analyzing dissolved copper content in electrical insulating oil using Atomic Absorption Spectrophotometry. Most laboratories (and equipment owners) are interested in more than just the dissolved copper content of the insulating oil, and many are using a method to detect and quantify a number of metals simultaneously using Inductively Coupled Plasma Spectrophotometry (ICP). The ICP has been used successfully for years to determine metal content of lubricating oils. A new ASTM standard method for ICP analysis of transformer oil is being written.

ICP determination can be done for a broad range of metals, but is most commonly run for dissolved iron, copper, and aluminum. The method involves injecting a test specimen of the oil into a radio frequency induced plasma flame at approximately 10,000 °C. The organic compounds of the oil are completely destroyed, leaving behind any metal atoms that were present. At these terribly high temperatures, the metal atoms are excited enough to emit specific wavelengths of light. For example, copper atoms emit light at a wavelength of 3247.5 angstroms. A detector records the intensity of each wavelength of light emitted by the metals, which indicates the amount of the specific metals present. Dissolved metals in oil can be detected by the ICP at levels of a fraction of one part per million. In our laboratory, the detection limits for copper, iron, and aluminum range from about 0.025 parts per million (25 parts per billion) to about 0.100 parts per million (100 parts per billion).

The profile of metals does not usually have a classification as to whether the results

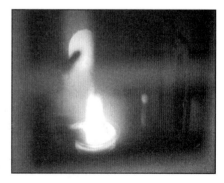

Figure 3.13 - ICP in operation.

are acceptable, questionable, or unacceptable. Each profile is interpreted according to the diagnostic value of the results. Most often, metals in oil analysis is performed to help further identify and provide the location for transformer fault conditions that were diagnosed earlier by dissolved gas analysis. Arcing, sparking, and severe hot spot overheating conditions may dissolve metals. These conditions also need to be found and corrected to reduce the risk of them worsening which may lead to an unplanned failure of the equipment. Dissolved metals analysis can be useful in helping to locate such faults. An increase of about 0.25 parts per million (250 parts per billion) over the previous baseline is usually considered to be significant.

Analysis for Furanic Compounds
ASTM Standard Method D 5837

The solid insulation in a transformer is made up of paper. Paper is made up of cellulose fibers. Cellulose is a polymer formed from glucose molecules. The following diagram illustrates the structure of the cellulose polymer (each ring is a glucose molecule):

(The letter "n" indicates that this is a long chain of undetermined and unspecified length.)

When the paper is brand new, before it has been installed in a transformer and factory dried, the average cellulose polymer chain is 1000 to 1200 glucose molecules long. Installation and drying breaks down the cellulose a little bit, so that new paper in a new transformer has slightly shorter polymer chains – about 800 to 1000 glucose molecules long. We call the average cellulose chain length the "Degree of Polymerization" (DP) of the paper. As the paper ages, there is a natural and gradual breakdown of the polymer chains. As the chains get shorter, the mechanical strength of the paper is also reduced.

A lot of work has been done relating the DP of the paper to the mechanical strength (tensile strength) of the paper. IEEE has even defined the end of reliable life for a transformer in terms of the DP of the paper that makes up the solid insulation. When DP has been reduced to 200, the paper is so weak that any stress will disrupt the paper and lead to failure. This is the practical definition of the end of reliable life for the solid insulation and therefore the end of life for the equipment.

When the cellulose chain breaks and two shorter chains are formed, the breakdown process "kicks out" one or more of the glucose molecules and also creates some water, carbon monoxide, and carbon dioxide. The glucose molecule changes chemically during this process and forms a compound containing a furan ring.

The compounds containing a furan ring are partially soluble in oil. They form in the paper, but partly migrate into the oil where we can detect them by chemical analysis. The temperatures at which the breakdown occurs and the presence of abnormally high levels of oxygen and moisture determine which compounds are formed. A number of these furan derivative (furanic) compounds can be formed, but some are much more common than others.

- **2-furaldehyde** is also called furfural or furfuraldehyde in some of the literature. This is the most commonly found furanic compound. It is formed by general overheating of the cellulose or may be present as the result of a past fault condition. It is the most stable of the furanic compounds under conditions found inside the transformer. The other compounds break down to form additional 2-furaldehyde.

- **2-furyl alcohol** is also called furfuryl alcohol or furfurol in some of the literature. This furanic compound forms in the presence of excessive moisture, and indicates an active condition of the paper breaking down due to high moisture levels in the solid insulation.

- **2-acetyl furan** is the most rare of the furanic compounds found in operating transformers. It is found most often in failed transformers that have been subjected to lightning strikes leading to a tentative conclusion that formation of 2-acetyl furan may be a result of high levels of electrical stress.

- **5-methyl-2-furaldehyde** forms as a result of localized and intense overheating of the cellulose and is an indication of an active condition involving a high temperature hot spot fault.

- **5-hydroxymethyl-2-furaldehyde** forms as a result of the paper breaking down in the presence of excessive amounts of oxygen and is an indication of an active condition involving oxidation of the solid insulation.

Structures of Furanic Compounds

2-furaldehyde 2FAL

5-methyl-2-furaldehyde 5M2F

2-acetyl furan 2ACF

2-furyl alcohol 2FOL

5-hydroxymethyl-2-furaldehyde 5H2F

The change in furan content – the amount of furanic compounds generated during the testing interval – is the most important parameter for determining whether there is an active fault in the equipment that needs to be addressed. For a first analysis, where there is no past history – or where the past history is so old as to be virtually meaningless – we use the following standards for interpreting results:

- **0 to 20 ppb total furans** - Background, this is essentially a new transformer.
- **21 to 100 ppb total furans** - Acceptable, this represents normal aging.
- **101 to 250 ppb total furans** - Questionable, this represents probable accelerated ageing.
- **251 ppb total furans and up** - Unacceptable, this represents significant accelerated ageing.

In addition to our AC, QU, and UN ranges, we consider very high levels to be of more immediate concern. Levels over 1000 ppb indicate severe, irreversible damage to the solid insulation. We consider this to be the start of the "danger zone" because we start to see transformer failures in the range of 1000 to 1500 ppb total furans. We typically do not recommend reclaiming or other oil maintenance procedures for transformers where the total furan content exceeds approximately 1000 ppb. End of reliable life for a North American built transformer that has aged gradually, without hot spots, is approximately 2800 ppb total furans. We typically recommend replacement of transformers that exceed 2500 ppb total furans.

The reason for using the 100 ppb and 250 ppb levels as cut-off points is that these levels are consistent with the rest of our oil classification system. We place the upper limits of our AC ranges to correspond with the end of the "sludge free" operation zone for oil quality parameters – this is the point where damage to the paper caused by oxidation products begins. 100 ppb of total furans corresponds to enough paper breakdown to be roughly equivalent to a 10% loss of life, based on the reduced strength of the paper.

Similarly, the end of the QU range is intended to be equivalent to a point where substantial and measurable damage to the solid insulation occurs. 250 ppb of total furans corresponds roughly to a 25% loss of life.

Degree of Polymerization
ASTM Standard Method D 4243

If a small sample of the insulation paper is obtained from the transformer, the degree of polymerization may be obtained by directly testing the paper sample. The sample is dissolved in a special solution, and the viscosity is measured by passing the dissolved paper through a small orifice. The viscosity measured in this way has a direct relationship to the degree of polymerization of the paper.

This is a more practical method of measuring the remaining life of the paper compared to tensile strength testing due to the much larger sample required for tensile strength determinations. There still are difficulties, however. The process requires that an outage for the equipment be taken and that the unit be opened and perhaps drained to obtain a sample. Where can one obtain a representative sample? There is no written standard practice available defining how and where such a sample can be taken. For a transformer that is intended to continue in service, DP testing is also impractical. The major uses of DP testing are to evaluate transformers being taken out of service or to define rewinding and remanufacturing needs.

Calculating Degree of Polymerization and Remaining Life from Furan Results

Degree of polymerization can be calculated from furan results. Since an oil sample is more practical to obtain and since the furan content gives a more average value for the calculated DP, the results are generally more useful and representative.

There are two distinct populations of transformers. Transformers that do not have thermally upgraded paper form a much higher concentration of furans compared to transformers that do have thermally upgraded paper. The furanic compounds are also somewhat more subject to decomposition in the presence of the additives used to thermally upgrade the paper.

Transformers without thermally upgraded paper typically included 55 °C rise transformers manufactured in North America before the early 1960's and almost all transformers manufactured outside of North America – only recently have manufacturers in Europe and Japan offered thermally upgraded paper, and it remains an option that the purchaser has to specifically request. 55/65 °C rise and 65 °C transformers manufactured in North America since the early 1960's typically used thermally upgraded paper (this includes almost all North American manufactured units in that time period).

Furan content is a reasonably good predictor of DP if one segregates these two populations and applies a different calculation for estimating DP to each population. Since DP has been established as a measure of end of reliable life, the furanic compounds analysis can be applied to that use, as well. For transformers without thermally upgraded paper, the best estimation of DP is calculated using the 2-furaldehyde content. For transformers with thermally upgraded paper, the calculation is based on the total furan content. The equations used for these calculations are updated periodically. The ones that are being used at press time can be summarized by the following table:

Not Thermally Upgraded Paper 2FAL (ppb)	Thermally Upgraded Paper Total Furans	Calculated DP	Estimated % Life Used
58	51	800	0
130	100	700	10
292	195	600	21
654	381	500	34
1464	745	400	50
1720	852	380	54
2021	974	360	58
2374	1113	340	62
2789	1273	320	66
3277	1455	300	71
3851	1664	280	76
4524	1902	260	81
5315	2175	240	87
6245	2487	220	93
7337	2843	200	100

Analysis of Polychlorinated Biphenyls (PCBs) in Insulating Oil
ASTM Standard Method D 4059

The ASTM standard method for PCBs in insulating oil uses gas chromatography with either a packed column or mega-bore capillary column and an electron capture detector. PCB content is quantified by comparison to standards prepared from Aroclors, commercial mixtures of PCBs that were used as insulating liquids, and reported as parts per million by weight (milligrams per kilogram) of the appropriate Aroclor(s). For classifying equipment for environmental management according to the Federal PCBs rules contained in 40CFR761, this method of analysis and reporting is preferred.

There are other PCB analysis methods, mainly those EPA methods from SW-846, that are used as appropriate for waste oils and solids, to characterize spill sites, and to confirm clean-up efforts. Some of these are Aroclor matching patterns while others quantify individual congeners of PCB.

The generally accepted limit of detection for method D 4059 is 2 ppm. Individual laboratories may establish a lower method detection limit, and should establish the precision for the method as run in their own laboratory. This is important, since the EPA has advised owners to take the recognized error in their test results into consideration when planning a PCB management strategy.

Particle Count and Distribution
ASTM Standard Method D 6786

Particle counting – both total count and the size distribution – is being used much more frequently for electrical insulating oils. A new ASTM standard, using automatic optical particle counters, was approved in 2003 and will be published in the 2004 ASTM Annual Book of Standards, Volume 10.03. The automatic counters view a test specimen of insulating liquid optically. They count and assign a size to a representative group of particles and then calculate the total number of each size in a given volume of oil and report the results as a distribution among several size ranges. Laboratories report the size distribution as particles per milliliter (frequently 10 mL or 100 mL) in each size range. As long as the units are clearly defined in the report, there is not much difference. Some prefer the particles per 100 mL reports, though, because they can report low concentrations of larger particles more precisely.

Interpretation standards have not yet been developed. Most laboratories that offer the test have developed interim internal standards for interpreting the results.

Some uses for particle count analysis are to support in-service transformer oil testing or for testing oil in new installations or after servicing. Particle count analysis can be used to identify potential problems with excessive insulation fibers or other solid suspended particles or can be interpreted with other oil monitoring tests to diagnose low dielectric breakdown voltage values using the D 1816 method. Particle count and distribution is also gaining wider use in testing other types of equipment, particularly load tap changers and oil filled circuit breakers.

Particle count distribution are frequently reported graphically, like this example:

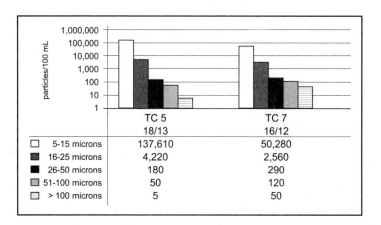

The above graphic is from the analysis of oil samples from two load tap changers. The ratio underneath the TC identifications represents the rating of the results according to ISO method 4406. The rating counts the total number of particles 5 microns and larger for the top number and the number larger than 15 microns for the bottom number. For a tap changer, the distribution for TC 5 above is elevated for both total particles 5 microns and larger and for particle larger than 15 microns. The overall distribution for TC 7 is more normal in this regard, but it is skewed abnormally toward particles larger than 25 microns.

Particles and Filming Compounds Analysis
Proprietary Methods

Frequently, the number and size of particles in an oil sample is important, but the types of particles present can provide more useful information. Also, the tendency of the oil to form additional solid materials and films can be evaluated.

This can be particularly useful to troubleshoot specific problems or as a monitoring test for equipment such as load tap changers. Excessive particle loading of LTCs may indicate the need for oil maintenance while the presence of some specific types of large particles, such as large ferrous arcing particles, may indicate operational problems or the need for maintenance of the contacts and mechanism.

TESTS	AC Results	QU Results	UN Results
Oil Screen:			
Neutralization Number	continue monitoring	hot oil clean	hot oil clean
Interfacial Tension	continue monitoring	hot oil clean	hot oil clean
Dielectric Breakdown	continue monitoring	investigate	investigate
Color	continue monitoring	investigate	investigate
Appearance	continue monitoring	investigate	investigate
Relative Density	continue monitoring	investigate	investigate
Liquid Power Factor	continue monitoring	investigate trends	investigate trends
Oxidation Inhibitor	continue monitoring	continue monitoring	reinhibit or hot oil clean
Moisture	continue monitoring	field repair/dehydrate	field repair/dehydrate

Table 3.3 - Action recommendations based on Oil Screen, Liquid Power Factor, Oxidation Inhibitor and Moisture Test results

MAINTENANCE

- Reinhibit (only when results except inhibitor are AC)

- Hot Oil Clean (Fullers Earth, Fluidex treatment)

- Dry out (dehydrate)
    - Factory or service shop dry out
    - Full vacuum field dehydration
    - Hot oil recirculation
    - Partial vacuum dry out
    - Passive on-line dehydrators (InsulDryer)

- Field Repairs (leak check/repair)

Table 3.4 - Key to Maintenance Recommendations

For referenced ASTM standards, visit the ASTM website, www.astm.org, or contact ASTM Customer Service at service@astm.org. For Annual Book of ASTM Standards volume information, refer to the standard's Document Summary page on the ASTM website.

Dissolved Gas In Oil Analysis (DGA) by Gas Chromatography (GC)

Introduction

Combustible gases are generated as a transformer undergoes thermal and electrical stresses over time. Both the oil and cellulosic insulating materials break down as a result of these stresses to yield gases.

The presence and quantity of these individual gases, extracted from the oil and analyzed, reveal the type and degree of the condition responsible for the gas generated.

The rate and amount of gas generated is important. Normal aging produces gases, but at an extremely slow rate. Incipient or newly formed fault conditions cause immediate and noticeable changes in the dissolved gas content of a transformer's oil. More importantly, an overwhelming majority of these incipient faults give early evidence of a pending failure or problem, and therefore can be detected when a transformer's oil is analyzed. Gas chromatography (GC) is the most practical method available to identify the combustible gases. GC involves both a qualitative and a quantitative analysis of gases dissolved in transformer oil. (Table 3.5)

History

The presence of dissolved combustible gases in transformer oil has been known since World War I. Laboratory work done at that time details a characteristic of hydrocarbons little appreciated – that gases radically different from oil vapor form on the passage of a disruptive discharge at or below the oil surface.

Typical Gases Generated by Transformer Faults[1]

Name	Symbol
Hydrogen[2]	H_2
Oxygen	O_2
Nitrogen	N_2
Methane[2]	CH_4
Carbon Monoxide[2]	CO
Ethane[2]	C_2H_6
Carbon Dioxide	CO_2
Ethylene[2]	C_2H_4
Acetylene[2]	C_2H_2

[1] Even though detectable and identifiable, the presence of propane (C_3H_8), propylene (C_3H_6), and butane (C_4H_{10}) are not used in diagnostic methodologies.

[2] Denotes combustible gas.

Table 3.5

Prior to 1920

Actual realization of the gassing problem in energized transformers was discovered and then documented in the February 1919 issue of *The Electric Journal:*

> "A further characteristic of hydrocarbons hither to (known) but little appreciated is the formation, on the passage of a disruptive discharge, at or below the oil surface, of gases radically different from oil vapor. The vapors arising from the oil surface as a result of a gradual increase in oil temperature and the accumulation of which forms the basis of the flash point test, are still essentially oil and subsequently condense, on decreasing the temperature, to resume their original character. However, the gases resulting from the disintegration of the oil molecules are permanent over ordinary temperature ranges and are present in the following approximate percentages:

Carbon Dioxide	1.17
Heavy Hydrocarbons	4.86
Oxygen	1.36
Carbon Monoxide	19.21
Hydrogen	59.10
Nitrogen	10.10
Methane	4.20
Total	**100.00**

This analysis indicates the presence of a preponderating quantity of hydrogen and a relatively small amount of hydrocarbons. The nitrogen and oxygen in the reaction products show the presence of 13% occluded air. In view of the violence accompanying the reunion of oxygen and hydrogen, as well as the fact that hydrogen is explosive in air, within a range of 10 to 66% it is evident that a destructive force can readily be obtained on the ignition of an atmosphere consisting of air and decomposed oil.

With a proportion of hydrogen to oxygen of two to one existing in the gaseous mixture, thereby assuring complete combination of both gases and most rapid reaction the speed of flame propagation is about 11,000 feet per second, or approximately ten times that of sound waves in air.

With gases initially at atmospheric pressure, an explosion would produce a pressure of eight atmospheres or about 120 pounds. Assuming the evolution of gases from an arc below the oil sufficiently rapid to create a pressure above the oil, in spheres, the maximum pressure of the explosion wave would be about 12 atmospheres or 180 pounds per square inch. As this wave, however, has a very steep front due to its high velocity, the force applied to the structure is in the nature of an impact and the stress on the parts above the oil level may approximate 360 pounds per square inch, representing a loading sufficient to distort the container or shear the holding bolts.

Since the temperature required for the disintegration of the oil is necessarily very great, slight differences in flash or fire points of the oils used do not materially influence the ease with which such disruption is produced. No appreciable protection is, therefore, obtained by substituting a high flash for a low flash oil. In fact it has been observed that, due to the lower vapor pressure and consequent decreased dilution of hydrogen by the less active hydrocarbon vapors, explosions in the higher flash oils were occasionally more violent."

The Buchholz Relay (1919)

In response to the acknowledged problem of gas formation by faults, power engineers have developed methods of collecting the gases generated in the transformer with expectation of being able to diagnose the source of the problem. The classical example of such a device was the Buchholz Relay introduced in 1919. This attachment would detect the passage of gas bubbles generated in the transformer as they moved through the pipe connecting the transformer to the conservator tank.

It created international interest in 1927 when a maintenance technician, using a match, lit the gas exiting a transformer vent tube. This interest eventually led to laboratory methods to identify and quantify the gases produced.

As a note of **caution:** this story illustrates how all transformers will have combustible gases in the gas space and need to be purged prior to maintenance.

Other devices that could collect or detect gas formation were developed. Once operation of these attachments indicated evidence of gas, the gases were sampled and then transported to a laboratory for analysis.

Field Gas Detectors

The inconvenience of depending on a laboratory for the analysis led H.H. Wagner in 1959 to develop a field combustible gas detector that could sample and identify the presence of combustibles in the head-space of a transformer. (Note: Not to be confused with ASTM D 3612, Test Method C also referred to as "Head Space Analysis.") This fault gas detector would give a measure of the total combustible gas present, expressed as a percent – without identifying the individual gases.

Gas Chromatography Invented

M.S. Tswett (Tsvet), a Russian botanist, was having trouble analyzing a mixture of plant pigments. In analytical chemistry, the most difficult step is oftentimes separating the mixture prior to identification. Dr. Tswett first investigated this art of separation in 1903.

The Russian packed a glass column completely with a powdered calcium carbonate mixture (blackboard chalk). He then poured an unknown liquid through the column where it stratified in different colored bands. The tube was broken and each separate band was analyzed and found to be a fairly pure compound. Chroma is the Greek word for "color" and thus the designation "chromatography" (1906).

Following the suggestion of A.J.P. Martin and R.L.M. Synge in a study for which they were later awarded the Nobel Prize, A.T. James and Martin introduced gas-liquid chromatography in 1952. Chromatography has since advanced to where it became possible to separate a mixture of gases. Therefore, today we have liquid chromatography and gas chromatography as useful tools. In the early 1960's, gas chromatography (GC) was first applied to the identification of fault gases dissolved in transformer oil. Other gas chromatographic methods utilizing electron capture detectors are employed to determine PCB content in insulating fluids. High Performance Liquid Chromatography (HPLC) is used in identifying furanic compounds.

Using gas chromatography for combustible gas identification has placed the problem back in the laboratory - where not only can the problem be quantified, but where we can often pinpoint the specific source.

Sampling

The first important step in performing dissolved gas analysis is securing a representative sample for analysis. A transformer technician or an electrician can usually draw a sample in less than five minutes (Figure 3.14). Care must be taken to employ good technique in drawing samples to prevent contamination. Extreme caution must be taken when sampling energized transformers. Your procedures should be reviewed and approved by your organization's safety officer. Current transformers (CT's) should never be sampled energized. Regardless of the sampling container type, certain precautions are required:

Sample a unit only under positive pressure. Don't try to draw a sample from a unit under negative pressure. Drawing an air bubble in through the bottom valve could cause failure.

Flush the sampling throughly and discard to insure that the sample is not contaminated. Protect the sample from sunlight. Forward to laboratory as soon as possible (prevent loss of dissolved gases).

ASTM D 3613 (Standard Practice for Sampling Insulating Liquids for Gas Analysis and Determination of Water Content) allows sampling into three approved types of containers. Table 3.6 summarizes pro and con views regarding each type used.

A known volume of oil (usually 40 ml) is drawn from the transformer and sent to a laboratory where the dissolved gases are extracted.

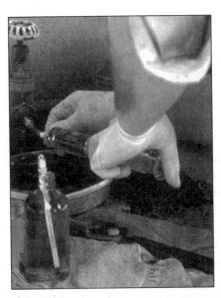

Figure 3.14 - A technician taking a syringe oil sample from a transformer's bottom filter press valve.

Comparing Typical GC Sampling Containers ASTM Standard D-3613

Type of Sample Container	Glass, Hypodermic Syringe	Stainless Steel Bottle	Flexible-sided Metal (tin) Can
Location Used	U.S. utilities and industry (gas blanketed and open breather units)	U.S. utilities, industry, and transformer manufacturers	Europe (transformer manufacturer) U.S. - some utilities
Pros	Inexpensive, Allow for volume change per temperature change[1]	Contamination-proof, leak-proof[1]	Allow for volume change per temperature change; larger container[1]
Cons	Breakable	Rigid construction; no visible inspection; can't see if a bubble is in the container	Limited use
Comments	Test within short time; periodic gas test suggested	Use for extended storage	Limited

[1]Each container may be cleaned

Table 3.6

Typically, a 50 mL glass syringe fitted with a three-way stopcock is placed in a special shipping carton (Figure 3.15). A Tygon tube adapter is provided so that the syringe can be filled from the bottom filter press valve of most units.

Each syringe used for sampling should be checked periodically for very small gas leaks.

The gas tightness of a syringe might be tested by storing an oil sample with known amounts of dissolved gases in the syringe for two weeks, and analyzing the content of hydrogen and methane at the beginning and end of that period. A syringe can be considered tight when the loss of hydrogen is less than 2.5% and loss of methane is less than 0.5%.

It follows then that such oil samples should be tested within a short period of time. Data has shown that storage for three additional weeks results in lower gas values. On the other hand, syringe samples do not appear to be affected significantly by ambient temperature cycling in transit.

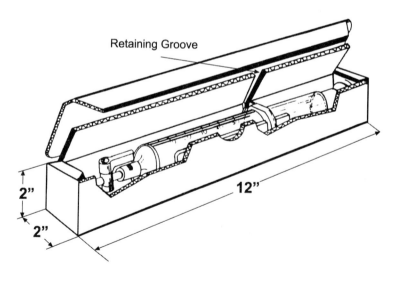

Figure 3.15 - A syringe, oil sample, and shipping carton.

Test Description

ASTM test method D 3612 describes three different extraction methods.

Method A uses a laboratory-based vacuum extraction procedure (Figure 3.16). This approach utilizes a glassware apparatus with a mercury piston to concentrate and return to atmospheric pressure the gases extracted at the top of the glass column.

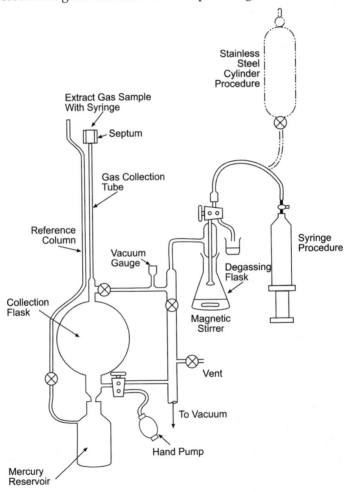

Figure 3.16 - Typical Details of Laboratory apparatus for extraction of gas from insulating oil.

Method B, known as a "stripper column extraction," is a direct injection method. It utilizes a multi-port set of valves in the gas chromatograph itself where the extraction is accomplished.

Method C, known as head space analysis, was approved in 2000. In this method a sample is placed in a vial where the dissolved gases equilibrate with the gas space above the sample.

Method A, which uses mercury, is described in further detail following. There are similar proprietary methods of vacuum extraction that use a mechanical device such as a metal piston in place of the mercury used to equalize pressures.

A typical sequence for gas extraction from insulating oil consists of these steps:

1. Injection of syringe oil sample into the extraction apparatus. (Figure 3.17) Connect syringe to the degassing flask. In performing this step, make sure any air bubble is removed and watch for foaming of the gas.

Figure 3.17 - Injection of syringe oil sample into the extraction apparatus.

2. Boiling of the gas in the gassing flask should last for about 15 minutes. (Figure 3.18) The gases release from the oil by vigorous stirring under high vacuum.

3. Normalization of pressures for standard temperature and pressure of 760 torr (101.325 kPa). A mercury "piston" is utilized

Figure 3.18 - Boiling of gas in the gassing flask.

to compress the gases to Standard Temperature and Pressure (STP).

4. Measurement (mL) of extracted gas in collection tube. The percentage of gas in the oil is calculated from the corrected volume of extracted gas. Figure 3.19 shows extraction of gas from the gas collection tube with a syringe for GC analysis.

5. The syringe of extracted gas is then injected into the gas chromatographic system as shown in Figure 3.20.

Figure 3.19 - The extraction of gas with a syringe for GC analysis.

Gas Analysis by Chromatography

A gas chromatograph is a means for separating gas components on a column and then detecting them (Figure 3.21) individually.

A typical gas chromatograph set-up (Figure 3.22) consists of a cylinder of carrier gas, a flow controller and pressure regulator, a sample injection port (sample inlet), adsorption column, detector (with necessary electronics), thermostats, a recorder, and a microprocessor for a data printout.

Figure 3.20 - Injection of gas sample into GC instrument.

Figure 3.21 - A typical gas chromatographic system.

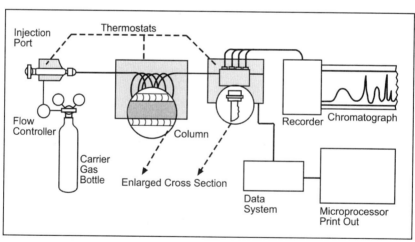

Figure 3.22 - Schematic drawing of gas chromatographic system.

The time required for this analysis ranges from 8 to 30 minutes. Accuracy attainable in the gas chromatograph depends upon the care exercised in obtaining the gas sample, the choice of detector, and daily calibration of the test instrument to assure reliable results.

The limits of detection are from 5 to 20 ppm for hydrogen, 1 ppm for hydrocarbon gases, and 2 to 25 ppm for oxides of carbon. Detection limits vary as a function of the type of gas (carrier gas), sample container used and method of extraction.

Data Reproducibility

A question that inevitably arises concerning any laboratory data is, "What is the reproducibility of the dissolved gas analysis between various laboratories?" Table 3.7 shows ASTM desired results with a range difference from 0.016 to 0.171%, and averaging less than 0.1% difference. Thus the data independently obtained by dissolved gas analysis from two qualified labs is precise and accurate. The data comes from a typical chromatogram (Figure 3.23). Nevertheless, interpreting the same data is clearly altogether another matter.

Dissolved Gas-in-Oil Analysis by Gas Chromatography Laboratory Reproducibility

Transformer Number	MVA Rating	Gallons of Oil (thousand)	Combustible Gas Content (CGC) as percent of Total Gas Content		Percent Difference Combustible Gas Content (CGC)
			S.D. Myers Lab	Second Laboratory	
1	45	13.5	0.409	0.425	0.016
2	45	18.7	2.258	2.285	0.027
3	45	9.8	0.504	0.463	0.041
4	292	24.0	0.520	0.685	0.165
5	292	24.0	0.903	0.732	0.171
6	308	23.7	0.477	0.395	0.082

Table 3.7

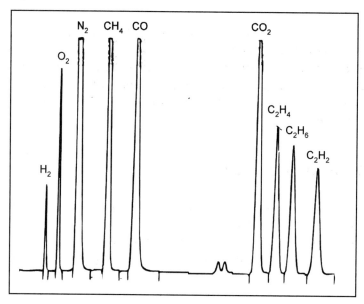

Figure 3.23 - Sample of printout from chromatograph.

The Physics of Gas in Oil

If a barrel of new oil were sampled and the gases extracted and analyzed, a typical analysis of the gas absorbed by the oil would show approximately 69.8 volume percent nitrogen and 30.2 volume percent oxygen. Other atmospheric gases (primarily CO_2) may be present in small amounts. Light hydrocarbon gases will usually not be present under normal conditions.

As a comparison, the atmosphere contains 79% nitrogen and 20% oxygen. The difference between nitrogen/oxygen content of oil and the atmosphere comes from different solubilities of the gases in oil. Nitrogen is 8.6% soluble in oil while oxygen is 16% soluble in oil (Table 3.9). Recalling that oxidation (made possible by the presence of oxygen) degrades both oil and paper insulation in transformers, we see that mother nature is working against us right from the start!

The benefits of low gas content should not be underestimated since gas bubbles which might be formed in the transformer will then be quickly absorbed before they combine and cause a breakdown. Gas supersaturation, with the danger of the formation of free gas bubbles on sudden pressure drops, is thus completely eliminated.

GC vs. Other Methods

The earliest methods of detecting combustible gases involved direct flammability tests or chemical analysis of the gases over the oil. However, these methods were not sensitive enough by today's standards, nor, more importantly, timely enough to be of value.

Five methods of fault gas detection are used at the present time. We will consider both advantages and disadvantages, and document how dissolved gas analysis has gained acceptance as the most informative method in today's computer-minded age. Table 3.8 summarizes the most pertinent facts.

Gas Collector Relay (Buchholz System)

This method has found limited application in the U.S. but extensive use in Europe. Recall that the generated combustible gases are highly soluble in oil. In a completely oil-filled transformer, small amounts of combustibles from an incipient fault may be almost totally absorbed in the oil before any appreciable free gas can accumulate in the collector relay. And when it does collect in the relay, you are now in possession of historical information; that is, your transformer has had a problem and you are now just discovering the evidence.

The Buchholz relay may respond to low energy partial discharges or local overheating. However, since the relay and collector relay are normally remote from the center of critical action the windings, and other key areas, it is impossible for the Buchholz

Method	Location	Required	Detected	Quantification	Application	Comments
Gas Collector Relay - Buchholz	Field - on the transformer	Gas actuates "fault alarm"	Measure total gas accumulated	No - screen test only	Conservator type units only	More popular in Canada and Europe
Total Combustible Gases (TCG)	Field	Gas sample of head-space	Estimated total combustible gases in head-space only	No - screen test only percent of total combustibles	Gas blanketed units	Results affected by gas solubilities and temperature
Gas Blanket Analysis, ASTM Sampling Method	Laboratory	Gas sample Positive Pressure	Identifies gases in head-space	Yes - parts per million (ppm)	Gas blanketed units	Sampling method ASTM D-2759
Dissolved Gas Analysis (DGA) ASTM D-3612	Laboratory	Oil sample	Identifies gas dissolved in oil where generated	Yes - parts per million (ppm)	All oil-filled transformers	Indicates the type of problem
Hydrogen in Oil	Portable field analyzer	Oil sample	Identifies hydrogen dissolved in oil	Yes - ppm screen test; abnormal - lab DGA follow-up	All oil-filled transformers	Good single indicator of incipient fault
Portable Dissolved Gas Monitors	Field	Oil sample	Some detect up to 8 gases	Yes - parts per million (ppm)	All oil-filled transformers	Can be wired to communicate with remote location

Table 3.8 - Comparing methods for detection of fault gases.

Solubility of Gases in Transformer Oil

Hydrogen[1]	H_2	7.0% by volume
Nitrogen	N_2	8.6% by volume
Carbon Monoxide[1]	CO	9.0% by volume
Oxygen	O_2	16.0% by volume
Methane[1]	CH_4	30.0% by volume
Carbon Dioxide	CO_2	120.0% by volume
Ethane[1]	C_2H_6	280.0% by volume
Ethylene[1]	C_2H_4	280.0% by volume
Acetylene[1]	C_2H_2	400.0% by volume

Static equilibrium at 101.325 kPA and 25°C

[1]Denotes combustible gas

Table 3.9

system to respond to early stages of trouble in the heart of the transformer. In fact, as E. Dörnenberg reasoned, if the transformer is full of oil, oil must be in the immediate vicinity of the location of the fault. Thus, looking for the combustible gases in the oil is more efficient, considering their solubilities, than looking for them in the head-space or in a relay. Also, diagnostic interpretation of gas content has been well established for gas in oil – not for free gas in the head space.

Fault Gas Detection

Fault gas detection, probably the most widely used field method in the U.S., measures total combustible gases (TCG). Its major advantages include its speed and adaptability to field use. In fact, it can be used to continuously monitor a transformer. However, it detects the combustible fault gases only, and is applicable only to those units having a gas-space above the oil. The major disadvantage of the TCG method is that it gives only a single value for the percentage of combustible gases, and therefore does not identify or quantify what gases are present. The results are subject to such factors as temperature and pressure, as well as solubility of the various gases of interest. As shown in Table 3.9 the solubilities vary at standard temperature and pressure.

William Henry's Law requires that subsequent analysis be taken at similar temperature and pressure, otherwise, any data would be meaningless. "The solubility of a gas in a liquid is directly proportional to the pressure of the gas above the liquid at a definite temperature." There may be little tendency for the combustibles to migrate into the gas-space if the concentration is low.

Even a small amount (as little as 2 ppm) of acetylene, for example, portends great problems! Acetylene is released with great reluctance from the oil with the results that analysis of the free gas in the air-space may be quite misleading in terms of detectable combustible gases.

Gas Blanket Analysis

A laboratory analysis of a gas sample in the transformer headspace identifies all the individual gas components, but only in the headspace. Thus, as with TCG, the gases must first diffuse into the gas blanket.

Dissolved Gas Analysis

The most informative method of fault gas detection is dissolved gas analysis (DGA). In this laboratory method an oil sample is taken from a transformer; the dissolved gases are then extracted, separated, identified, and quantitatively determined.

Various lab methods have been used, including infrared absorption and mass spectroscopy, but gas chromatography has emerged as the most popular technique. Primary criteria include sensitivity required, capital cost, facility of operation, and so forth.

Diffusion of gases between liquid and gaseous spaces takes time, during which serious equipment damage can occur undetected, if only gas samples from the transformer head-space are analyzed for combustibles (methods number 1, number 2, and number 3, Table 3.8).

Monitoring the oil for dissolved gases offers both the required sensitivity, while giving the earliest possible detection of a newly-formed fault. The only disadvantage for DGA lies in that it can't be done readily in the field. On the other hand, this method is not only applicable to all types of oil-filled equipment, it gives the information required to properly evaluate the transformer's ability to live up to its intended function.

Test Data Interpretation

Introduction

The primary causes of fault gases are the thermal, electrical, and to a lesser degree mechanical stresses which result in the following conditions:

1. Partial Discharge (Corona), Sparking and Arcing

   a. Partial Discharge (Corona) - electrical stressing resulting in ionization; first occurs around 10,000 volts over sharp edges of current-carrying conductors.
   b. Sparking - a single, short electrical discharge lasting a microsecond or less.
   c. Arcing - a prolonged electrical discharge producing a bright flame-colored arc in contrast to dim corona-type glow.

2. Thermal Heating (hot spots and general over heating)

   a. Hot spots - localized overheating. Incipient faults can reach 500 °C without sufficient heat to char the cellulose.
   b. General overheating - without hot spots.

These symptoms differ essentially in the intensity of energy that is dissipated by the fault.

The different causes of gases accumulating in the oil include:

- Leak to atmosphere
- Nitrogen blanket contaminated
- Old fault corrected - not degassed
- FOA (OFAN or OFAF) transformer - motorburned out

- Hydrolysis (if free water is present)
- Transformer shipped with CO_2 gas
- Gases remaining from leak testing
- Gases from earlier faults leaching out of the solid insulation following partial rewind.

Combustible and other gases detected in the oil from these sources are indicative of abnormal conditions but not fault conditions.

Laboratory Demonstration - Origin of Combustible Gases

If we were to take a transformer's oil and paper insulation system to the laboratory to conduct experiments on models that stress the components similar to those conditions found in an operating transformer, we would be able to collect the following data:

1. Cellulose

    a. Overheating of cellulose - When cellulose is overheated in a closed system and the gases collected, analysis shows the primary products of decomposition at temperatures as low as 140 °C to be:

    - Carbon monoxide
    - Carbon dioxide
    - Water

    b. Pyrolysis of cellulose - When cellulose is heated to destruction (pyrolysis), Figure 3.24 illustrates the normal decomposition products. At temperatures above 250 °C (in a sealed unit) more CO than CO_2 may be formed, perhaps as much as four times CO than CO_2 by volume.

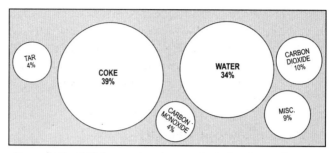

Figure 3.24 - Cellulose undergoing pyrolysis (heating to destruction).

2. Transformer Oil

   a. Overheating Oil - When insulating mineral oil is overheated at temperatures up to 500 °C, these hydrocarbon vapors will be liberated:

      - Ethylene
      - Ethane
      - Methane

      Other products liberated below 500 °C with some oxygen present are:

      - Carbon dioxide (400 °C)
      - Water (200 °C)

   b. Pyrolysis of oil - When oil is subjected to extreme electrical stress (such as an electrical arc), this quantity of gases will be liberated:

Hydrogen	60.0 - 80.0%
Acetylene	10.0 - 25.0%
Methane	1.5 - 3.5 %
Ethylene	1.0 - 2.9 %

Note the absence of carbon oxides in this case. All kinds of gases and higher molecular weight hydrocarbons are generated in the oil. It is very difficult to isolate and identify different liquid hydrocarbons produced by events in the life of a transformer. However, it is relatively easy to isolate and identify the gases.

Comparative Rates of Gas Evolution

As we continue to build our case of predicting incipient faults, look at Figure 3.25 which shows the relationship of decomposition temperature versus the type of gas generated in a fault condition. Thus, if one combines the solubility of the gas in the oil, Table 3.9, with the temperature it takes to generate the various gases, it can readily be seen that different faults will produce different evidences of their existence.

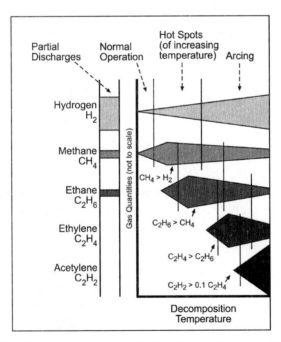

Figure 3.25 - Comparative rates of evolution of gases from oil as a function of decomposition energy. R.R. Rogers, "IEEE and IEC codes to Interpret Incipient Faults on Transfomers Using Gas-In-Oil Analysis."

Qualitative and Quantitative Interpretation

A number of methods are in use for the interpretation of the dissolved gas analysis (Table 3.10).

A generally accepted list of gases and associated conditions is found in Table 3.11.

Comparing Methods of Interpreting Dissolved Gas-in-Oil Data

Interpretive Method	Value of Method	Source	Comment
Detected Gases	Types of probable faults	ANSI/IEEE C57.104 Draft 10c 2003	Qualitative (Table 3.9)
Detected Gases	Types of faults	IEC 60599-1999	Qualitative (Table 3.9)
Key Gas	Type of general problem	Doble Engineering	Qualitative (Figure 3.20)
Key Components	Type of general problem	E. Dörnenberg	Qualitative; all components must be identified
Dörnenberg Ratios	Type of general problem	E. Dörnenberg	Qualitative (Figure 3.21)
Total Combustible Gases	Severity of problem	Doble Engineering	500 or more ppm - retesting is critical - establish a trend
Rogers Ratios	Severity of problem	R. R. Rogers	No mention of CO, simultaneous faults, ambiguous anaylsis, concentrations important (Table 3.11)
Combustible Concentration Limits	Guidelines only	E. Dörnenberg	Table 3.19
90% Norms Concentration	Guidelines only	ANSI/IEEE C57.104 Draft 10c 2003	Concentrations important; avoid legalistic application (Table 3.20)
90% Norms Concentration	Guidelines only	IEC 60599	

Table 3.10

Types of Probable Faults

Detected Gases	Interpretations
Nitrogen plus 5% or less oxygen.	Normal operation of sealed transformer.
Nitrogen plus more than 5% oxygen.	Check for tightness of sealed transformer.
Nitrogen, carbon dioxide, carbon monoxide, or all.	Transformer overloaded or operating hot, causing some cellulose breakdown. Check operating conditions.*
Nitrogen and hydrogen	Partial discharge, electrolysis of water, or rusting.
Nitrogen, hydrogen, carbon dioxide, and carbon monoxide.	Partial discharge involving cellulose or severe overloading of transformer.*
Nitrogen, hydrogen, methane with small amounts of ethane, and ethylene.	Sparking or other minor fault causing some breakdown of oil
Nitrogen, hydrogen, methane, with carbon dioxide, and small amounts of other hydrocarbons; acetylene is usually not present.	Sparking or other minor fault in presence of cellulose.*
Nitrogen with high hydrogen and other hydrocarbons including acetylene.	High energy arc causing rapid deterioration of oil.
Nitrogen with high hydrogen, methane, high ethylene and some acetylene.	High temperature arcing of oil but in a confined area, poor connections or turn-to-turn shorts are examples.
Same as above, except carbon dioxide and carbon monoxide are present.	Same as above, except arcing in combination with cellulose.*

*Check for furanic compounds.

Table 3.11

1. Key Gases

    Table 3.11 may be simplified by looking for "Key Gases" in the data and diagnosing the associated condition (Figure 3.26).

2. Key Components Technique

    Presence of certain components and their relative concentration gave rise to this Dörnenberg technique.

    Thermal decomposition (hot spots) - mainly C_2H_4, CH_4; less C_2H_6, H_2, sometimes C_2H_2 (reference Table 3.5).

    High intensity discharges - mainly H_2, C_2H_2, then CH_4, C_2H_4, and C_2H_6.

    Low intensity partial discharge (corona) mainly H_2, smaller amounts CH_4, C_2H_6.

    Cellulosic insulation decomposition - CO and CO_2.

    Once the type of fault is diagnosed, the severity of the problem becomes the next item of concern.

3. The Amounts of Key Gases

    Since the detection limits of the GC instrument producing the data are in the parts per million (ppm) range, schemes are being used to indicate the normal or abnormal levels of the individual gases.

4. Total Combustible Gases

    Combustible gases in the range of 0-700 ppm are normally an indication of satisfactory operation of the transformer (for an exception, see Table 3.18).

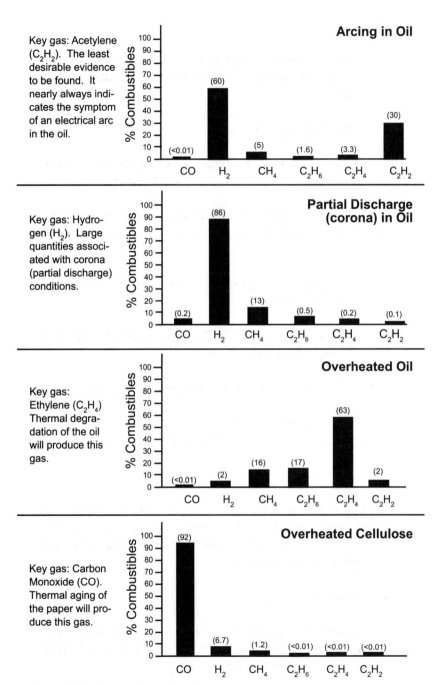

Figure 3.26 - Key gas diagnostic criteria

A range of 700-1900 ppm of combustible gas indicates that decomposition may be in excess of normal aging (suggesting a need for more frequent analysis).

More than 1900 ppm of combustibles usually means that decomposition is significant. The next step here should be establishing a trend.*

Accumulation of more than 2500 ppm of combustibles indicates substantial decomposition. The integrity of the insulation system is in question. Monitoring is essential. The transformer should be inspected for fitness, continued operation, or for repair. Refer to Draft 10c IEEE C57.104 (2003).

Ratio Analysis Methods

A second type of interpretation concerns the Dörnenberg and Rogers Ratio methods.

1. Dörnenberg Ratio Method

   One of the earliest methods is that of Dörnenberg in which two ratios of gases are plotted on log-log axes (Figure 3.27).

   The area in which the point falls indicates the type of fault that has developed.

2. Rogers Ratio Method

   This method has probably received the most publicity. Statistical study of nearly 10,000 gas analyses (United Kingdom utilities) suggested that certain types of fault con-

---

*If the amount of combustibles remains constant, then a possible self-healing effect may have taken place. If the amount increases, then the unit can be in the "danger" zone.

ditions could be differentiated with more detailed ranges and combinations of gas ratios. This has been confirmed by internal examination of a number of suspect transformers together with post-mortems on faulty units.

3. Duval Triangle (see Figure 3.28)

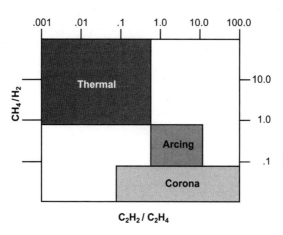

Figure 3.27 - Early Dörnenberg ratio plot of symptomatic indicators of faults

Suggested Diagnosis from Gas Ratios - Rogers Ratio Method

R2 $\dfrac{C_2H_2}{C_2H_4}$	R1 $\dfrac{CH_4}{H_2}$	R5 $\dfrac{C_2H_4}{C_2H_6}$	Suggested Fault Diagnosis
< 0.1	> 0.1 and < 1.0	< 0.1	Normal
< 0.1	< 0.1	< 0.1	Partial discharge (corona) - low energy density arcing
> 0.1 and < 3.0	> 0.1 and < 1.0	> 3.0	Arcing - high energy discharges
< 0.1	> 0.1 and < 1.0	> 1.0 and < 3.0	Low temperature thermal overheating
< 0.1	> 1.0	> 1.0 and < 3.0	High temperature thermal - less than 700°C
< 0.1	> 1.0	> 3.0	High temperature thermal - more than 700°C

Table 3.12

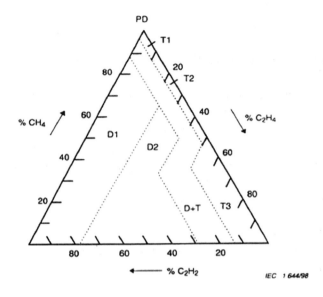

$$\% \, C_2H_2 = \frac{100\,x}{x+y+z} \quad \text{for } x = [C_2H_2] \text{ in microlitres per litre}$$

$$\% \, C_2H_4 = \frac{100\,y}{x+y+z} \quad \text{for } y = [C_2H_4] \text{ in microlitres per litre}$$

$$\% \, CH_4 = \frac{100\,z}{x+y+z} \quad \text{for } z = [CH_4] \text{ in microlitres per litre}$$

Key
- PD Partial discharges
- D1 Discharges of low energy
- D2 Discharges of high energy
- T1 Thermal fault, $t < 300\,°C$
- T2 Thermal fault, $300\,°C < t < 700\,°C$
- T3 Thermal fault, $t > 700\,°C$

Figure 3.28 - Duval Trangle - This was developed from a database of over 10000 transformers. The 3 gases must equal 100%, then plot amount on paper axis. The area where the lines meet is the discharge or thermal diagnosis as seen in the key. (Courtesy of Michel Duval)

Case Histories Illustrating Basic Transformer Faults

Even with a computer printout and various textbook schemes of data interpretation, the decision as to "When should a transformer be removed from service?" cannot be precisely defined for all instances. The final decision is therefore largely based on comparative field case histories that give data, diagnosis, and actual internal findings for a comparable situation. This data should also be compared with a similar normal operation's data.

Needless to say, the first table (Table 3.13) illustrates typical data from normal operation (in this case, a 25 MVA furnace transformer).

Tables 3.14 through 3.20 illustrate the four basic transformer faults and should be self-explanatory.

When applying a method of interpretation in the diagnostic phase of analysis, trends are the first thing to look for. Note in Table 3.15 the numbers in themselves are relatively small, but when the trend and the rate of generation of carbon monoxide and acetylene have a 300% and 600% increase respectively in less than six months, an unplanned shutdown (failure) is inevitable. Consider then the types of gases and volume of oil in which they may dissolve.

Unless the fault is very obvious, these considerations will act as a guide for specific actions, such as shutdown or continued analysis to determine the progression of the fault.

A Case History Illustrating Thermal Overheating of Transformer Oil

SDMYERS
SUBSTATION SOLUTIONS

Customer - ABC COMPANY
MFG - G.E. KVA - 25,000 TYPE - TRANSFORMER
S/N - G-8581627-A GAL - 5,595 LOCATION - FURNACE

GAS-IN-OIL Analysis Gas Chromatography Expressed in PPM

Date	H_2	O_2	N_2	CH_4	CO	C_2H_6	CO_2	C_2H_4	C_2H_2	Total Combustible	Total Gas
12/22/00	ND	12,408	68,570	Trace	101	ND	836	ND	ND	101	84,915

RECOMMENDATION: Retest in 1 year

Diagnosis: Operation is normal. In 12 samplings over the last 12 months the combustibles have averaged from a low of 101 to a high of 174 ppm.

Table 3.13

H_2	Hydrogen
O_2	Oxygen
N_2	Nitrogen
CH_4	Methane
CO	Carbon Monoxide
C_2H_6	Ethane
CO_2	Carbon Dioxide
C_2H_4	Ethylene
C_2H_2	Acetylene
ND	None Detected

A Case History Illustrating Thermal Overheating of Transformer Oil

SDMYERS
SUBSTATION SOLUTIONS

Customer - ABC COMPANY
MFG - S.D. Myers   KVA - 5,000   SUB NAME - MAIN   UNIT # - MN# 1
S/N - 123456   GAL - 1,170   LOCATION - OUTDOOR/GROUND

GAS-IN-OIL Analysis Gas Chromatography Expressed in PPM

Date	H_2	O_2	N_2	CH_4	CO	C_2H_6	CO_2	C_2H_4	C_2H_2	Total Combustible	Total Gas
12/29/00	2004	2799	66,637	9739	1737	2750	9596	5113	ND	21,343	100,375

RECOMMENDATION: Investigate

Diagnosis: The Key Gas, ethylene, indicates the overheating of the transformer oil in the 350°C - 400°C range, possibly at a bad lead or loose connection in the termal board or tap changer areas.

Findings: The iron bolts and nuts connecting the copper primary leads to the aluminum were carbonized and melted.

Solution: Repair and use brass nuts and bolts.

Table 3.14

H_2	Hydrogen
O_2	Oxygen
N_2	Nitrogen
CH_4	Methane
CO	Carbon Monoxide
C_2H_6	Ethane
CO_2	Carbon Dioxide
C_2H_4	Ethylene
C_2H_2	Acetylene
ND	None Detected

A Case History Illustrating Thermal Overheating of Transformer Cellulosic Insulation

SD MYERS
SUBSTATION SOLUTIONS

Customer - ABC COMPANY
MFG - S.D. Myers   KVA - 292,000   TYPE - TRANSFORMER
S/N - 6598783   GAL - 23,670   LOCATION - CITY

GAS-IN-OIL Analysis Gas Chromatography Expressed in PPM

Date	H_2	O_2	N_2	CH_4	CO	C_2H_6	CO_2	C_2H_4	C_2H_2	Total Combustible	Total Gas
12/11/99	57	2615	13,407	13	145	Trace	241	11	12	238[1]	16,501
12/14/99	50	3540	15,300	10	133	2	290	10	12	217	19,347
12/29/99	ND	5901	22,134	ND	8	ND	96	ND	ND	8[2]	28,139
01/08/00	ND	10,697	38,947	ND	Trace	ND	109	ND	ND	Trace[3]	47,948
06/30/00	ND	9046	35,314	Trace	57	ND	270	ND	ND	57[4]	44,687

RECOMMENDATION: Investigate

[1] Diagnosis: Minor but significant increases in carbon monoxide and acetylene. The influx of these gases is normally indicative of an arc involving the celluslosic insulation.

Findings: A flashover from the end shield to the corona ring found on one bushing.
[2] Oil degassed
[3] Too short of interval
[4] Unit operating normally

Note: New transformer, so gas levels are of concern.

Table 3.15

H_2	Hydrogen
O_2	Oxygen
N_2	Nitrogen
CH_4	Methane
CO	Carbon Monoxide
C_2H_6	Ethane
CO_2	Carbon Dioxide
C_2H_4	Ethylene
C_2H_2	Acetylene
ND	None Detected

A Case History Illustrating Transformer Arcing

SD MYERS
SUBSTATION SOLUTIONS
ACTS 4:12

Customer - ABC COMPANY
MFG - STANDARD   KVA - 20,000   TYPE - TRANSFORMER
S/N - RIL-1717   GAL - 4,095   LOCATION - CITY SUB

GAS-IN-OIL Analysis Gas Chromatography Expressed in PPM

Date	H_2	O_2	N_2	CH_4	CO	C_2H_6	CO_2	C_2H_4	C_2H_2	Total Combustible	Total Gas
02/21/00	127	6215	71,620	107	174	11	973	154	224	79,605	797[1]

RECOMMENDATION: Investigate

[1] Diagnosis: The Key Gas, acetylene, indicates an arcing condition with some cellulose involved.

Findings: The unit has suffered a through-fault. When the secondary windings moved up there was an arc from the secondary to the core.

The unit is being repaired.

Table 3.16

H_2	Hydrogen
O_2	Oxygen
N_2	Nitrogen
CH_4	Methane
CO	Carbon Monoxide
C_2H_6	Ethane
CO_2	Carbon Dioxide
C_2H_4	Ethylene
C_2H_2	Acetylene
ND	None Detected

A Case History Illustrating Partial Discharge in Transformer Oil

SDMYERS
SUBSTATION SOLUTIONS

Customer - ABC COMPANY
MFG - DELTA STAR   KVA - 1,500   TYPE - TRANSFORMER
S/N - W-208106   GAL - 510   LOCATION - MN PLANT SUB

GAS-IN-OIL Analysis Gas Chromatography Expressed in PPM

Date	H_2	O_2	N_2	CH_4	CO	C_2H_6	CO_2	C_2H_4	C_2H_2	Total Combustible	Total Gas
04/19/00	35,882	1201	16,043	8,470	97	2139	1290	7	2	46,597[1]	65,131

RECOMMENDATION: Investigate

[1] Diagnosis: The Key Gas, hydrogen, indicates (low energy partial discharge) with no cellulose involved.

Findings: Design problem with unit. Unit will continue to operate until it can be replaced.

Table 3.17

H_2	Hydrogen
O_2	Oxygen
N_2	Nitrogen
CH_4	Methane
CO	Carbon Monoxide
C_2H_6	Ethane
CO_2	Carbon Dioxide
C_2H_4	Ethylene
C_2H_2	Acetylene
ND	None Detected

A Case History Illustrating Transformer Arcing

SDMYERS ACTS 4:12
SUBSTATION SOLUTIONS

Customer - ABC COMPANY
MFG - WESTINGHOUSE   KVA - 13,500   TYPE - TRANSFORMER
S/N - 6995019   GAL - 6,513   LOCATION - SUBSTATION

GAS-IN-OIL Analysis Gas Chromatography Expressed in PPM

Date	H_2	O_2	N_2	CH_4	CO	C_2H_6	CO_2	C_2H_4	C_2H_2	Total Combustible	Total Gas
11/13/99	35	9250	81,154	25	71	ND	170	23	22	176[1]	90,750
11/16/99	ND	5162	22,301	1	1	ND	35	Trace	Trace	2[2]	27,500
02/28/00	ND	12,059	68,122	2	4	2	186	8	ND	16[3]	80,383

RECOMMENDATION: Investigate

[1] Diagnosis: Key Gas of acetylene and carbon monoxide indicate there is arcing involving cellulose.

Findings: Client reported the unit developed an arc from high voltage windings to the tank.

Special note: A transformer is normally considered as operating satisfactorily below 700 ppm combustibles.

[2] The unit had its oil degassed after the repairs were completed on the transformer.
[3] Operation after three months appear to be normal.

Table 3.18

H_2	Hydrogen
O_2	Oxygen
N_2	Nitrogen
CH_4	Methane
CO	Carbon Monoxide
C_2H_6	Ethane
CO_2	Carbon Dioxide
C_2H_4	Ethylene
C_2H_2	Acetylene
ND	None Detected

A Case History Illustrating Preventative Maintenance

SDMYERS
SUBSTATION SOLUTIONS

Customer - ABC COMPANY
MFG - GE   KVA - 10,500   TYPE - TRANSFORMER
S/N - H-878597-C   GAL - 2,030   LOCATION - CITY SUBSTATION

GAS-IN-OIL Analysis Gas Chromatography Expressed in PPM

Date	H_2	O_2	N_2	CH_4	CO	C_2H_6	CO_2	C_2H_4	C_2H_2	Total Combustible	Total Gas
08/17/00	168	7217	55,884	1,353	67	581	1,386	3,281	63	5513	70,000

RECOMMENDATION: Investigate

[1] Diagnosis: The Key Gas, ethylene, with methane and acetylene indicates a bad conductor joint or loose connection.

Findings: A loose connection was found in the thermal board. This was due to a cross-threaded nut that was not properly tightened at manufacturing.

Note: Low Carbon Monoxide (CO), therefore the following should be observed.

1. No paper involvement
2. Check metals to see if any are dissolved.

Table 3.19

H_2	Hydrogen
O_2	Oxygen
N_2	Nitrogen
CH_4	Methane
CO	Carbon Monoxide
C_2H_6	Ethane
CO_2	Carbon Dioxide
C_2H_4	Ethylene
C_2H_2	Acetylene
ND	None Detected

A Case History Illustrating Corrective Maintenance

SDMYERS
SUBSTATION SOLUTIONS

Customer - ABC COMPANY
MFG - GE   KVA - 13,470   TYPE - TRANSFORMER
S/N - D-583393   GAL - 1,570   LOCATION - SUBSTATION

GAS-IN-OIL Analysis Gas Chromatography Expressed in PPM

Date	H_2	O_2	N_2	CH_4	CO	C_2H_6	CO_2	C_2H_4	C_2H_2	Total Combustible	Total Gas
08/05/99	9817	2488	76,772	36,962	837	11,608	6,649	62,815	2,624	124,663[1]	10,572
08/30/00	23	28,365	90,420	506	Trace	2,229	2,972	7,836	213	10,807[2]	32,564

RECOMMENDATION: Investigate

[1] Arcing odor detected in oil. Unit was tested and found to have extremely high combustible gas content in the oil.

Diagnosis: The Key Gas, acetylene, with large amounts of ethylene, indicates extreme overheating of oil.

Findings: Problem found to be in tap changer. An inner-ring contact was badly burned, as was the stationary contact.

[2] Combustible gas problem (though considerably reduced) still exists. Retest one month to note how fast the gases are generating.

Recommendation: As combustible gas level remains high- repair and degas.

Table 3.20

H_2	Hydrogen
O_2	Oxygen
N_2	Nitrogen
CH_4	Methane
CO	Carbon Monoxide
C_2H_6	Ethane
CO_2	Carbon Dioxide
C_2H_4	Ethylene
C_2H_2	Acetylene
ND	None Detected

Cautions in Analyzing GC Data

Caution is necessary when you apply a method of analysis to your GC data. Dörnenberg and others published concentration limits (Table 3.21). They suggested that even though these gases were present in the stated quantities, the transformer was thought to be operating perfectly.

These suggestions were refined and then incorporated into the ANSI/IEEE Guide C57.104-1991 (Table 3.22) as "90% probability norms." This means that if the concentrations are exceeded, the equipment merits further investigation.

A second case history (refer back to Table 3.18) illustrates how every man-made rule has an exception. In this case, this unit was not operating normally, even though the combustible gas content was substantially below 700 ppm (176 ppm). Fortunately, noting the rapid increase in acetylene indicated the possibility of a problem.

Further caution is necessary when you leave textbook cases and apply a method of analysis to your data. The ANSI/IEEE document C57.104-2003 Draft 10c and IEC Standard 60599 1999 should be used as guides. If a legalistic method of viewing the

Concentration Limits

Hydrogen	H_2	100 ppm (V/V)
Methane	CH_4	120 ppm (V/V)
Ethane	C_2H_6	35 ppm (V/V)
Ethylene	C_2H_4	50 ppm (V/V)
Acetylene	C_2H_2	65 ppm (V/V)
Carbon monoxide	CO	3500 ppm (V/V)
Carbon dioxide	CO_2	2500 ppm (V/V)
Total Combustible Gas		**720 ppm (V/V)**

One ppm (V/V) of a gas denotes that 10^{-6} liters (or 1mm^3) of this gas dissolved in 1 liter of insulating oil at a pressure of 1 kgf/cm^3.

Table 3.21

data is used, someone will demand that at the norm value plus one ppm the transformer must be shut down. This action is not usually necessary and is certainly not what the guide intended.

Synopsis of the Procedure for Statistical Development of Surveillance Range Norms from a Gas Database

For an aid in completing the statistical steps in this procedure, see Duncan, A.J., Quality Control and Industrial Statistics, Revised Edition, R.D. Irwin Publisher. These steps can be done by hand, or with the aid of statistical and/or graphing software.

1. Form a database of a statistically sufficient number of transformers. These should be transformers without any known internal faults and of similar design and service.

2. Tabulate the data for each individual combustible gas (H_2, CH_4, C_2H_2, C_2H_4, C_2H_6, and CO) and the total combustible gas by listing them in decreasing order.

3. Compute the relative cumulative frequency for each individual value for each gas.

4. Then for each gas, plot the relative cumulative frequencies versus the corresponding individual gas values.

5. Draw a smooth curve through the points on the graph.

6. Then use the curve for each gas to obtain the 90th percentile value of the data for each gas. This is the 90% probability norm of the particular gas from your database and it signifies that 90% of the (normal) transformers in this database have values at or below the 90% norm.

7. Next construct a table similar to Table 3.22 for your own database. The calculated 90% norms will comprise the first borders (between normal and caution). The second borders

(between caution and warning) are scaled up from the first borders in the same proportion as they are in Table 3.22.

This method of deriving norms from a specific population of transformers will give values that are statistically representative of that population. Such values will be preferable to the survey values in Table 3.22. Periodically repeating this analysis as the database is updated and grows will refine the estimated norms and increase their validity.

Comparing Current Methods of Dissolved Gas Analysis Interpretation

As would be expected, differences of opinion exist concerning the preferred method of fault identification. The two methods used most often are the Key Gas and Rogers Ratio techniques. The best procedure appears to be the Key Gas method, with Rogers Ratios or Duval's Triangle used as a back-up or a confirmation. The IEEE and IEC Standards present tables of limiting value of individual gases.

Surveillance Range Guidelines for Type 1 Transformers with No Previous Combustible Gas Tests

Gases Generated	Surveillance Range	Normal	Caution	Warning[2]
Hydrogen	H_2	< 100	100 - 700	> 700
Methane	CH_4	< 120	120 - 400	> 400
Acetylene	C_2H_2	< 2	2 - 5	> 5
Ethylene	C_2H_4	< 50	50 - 100	> 100
Ethane	C_2H_6	< 65	65 - 100	> 100
Carbon Monoxide	CO	< 350	350 - 570	> 570
Total[1]		< 700	700 - 1900	> 1900

[1] TDCG = Total of all combustible gases

[2] Gas concentrations in warning range indicate a severe problem generally requiring immediate intervention or removal.

Table 3.22

(Table 3.23) It is clearly obvious at this time that interpretation of dissolved gas analysis requires both experience and judgement, regardless of the qualitative and quantitative tools available.

However, consideration should be given to the transformer type (whether it is conservator or non-conservator). As an important bit of information, the ratio method was devised for use on conservator type units (relatively common overseas); whereas in the U.S., the majority of samples used in S.D. Myers Inc. evaluations are from either sealed or gas-blanketed transformers. In addition, look for a change in ppm if you have a bench-mark value.

Actions based on TDCG (C57.104-1991)
Sampling Intervals and Operating Procedures for Corresponding Gas Generation Rates

	TDCG levels (ppm)	TDCG rates (ppm/day)	Sampling Interval	Operating Procedures
Condition 4	> 4630	> 30	Daily	Consider removal of service.
		10 - 30	Daily	Advise manufacturer.
		< 10	Weekly	Exercise extreme caution. Analyze for individual gases. Plan outage. Advise manufacturer.
Condition 3	1921 - 4630	> 30	Weekly	Exercise extreme caution. Plan outage.
		10 - 30	Weekly	Analyze for individual gases.
		< 10	Monthly	Advise manufacturer.
Condition 2	721 - 1920	> 30	Monthly	Exercise extreme caution. Plan outage.
		10 - 30	Monthly	Analyze for individual gases.
		< 10	Quarterly	Advise manufacturer.
Condition 1	≤ 720	> 30	Monthly	Exercise caution. Analyze for individual gases. Determine load dependence.
		10 - 30	Quarterly	Exercise caution. Analyze for individual gases. Determine load dependence.
		< 10	Annually	Continue a normal operation.

Table 3.23

Our own study comparing the numerical values of the ratios with the suggested limits provides a suggested diagnosis of the source. We have found that correlation between the two methods ranges from poor to good, depending upon possibly the incipient fault and the ratio method used. (Table 3.24).

Our lab experience since the mid 1970's continues to show an excellent record of predicting newly-formed faults in oil-filled transformers.

Rogers Ratio Method Applied to Case Histories

Case History Reference	$\dfrac{CH_4}{H_2}$	$\dfrac{C_2H_4}{C_2H_6}$	$\dfrac{C_2H_2}{C_2H_4}$
Case from Table x.13[1]	$\dfrac{8470}{25,882} = 0.23$	$\dfrac{7}{2139} = 0.003$	$\dfrac{2}{7} = 0.28$
Case from Table x.10[2]	$\dfrac{9739}{2004} = 4.9$	$\dfrac{5113}{2750} = 1.9$	$\dfrac{ND}{5113} = 0.0$
Case from Table x.16[3]	$\dfrac{25}{35} = 0.7$	$\dfrac{23}{0} = 0.0$	$\dfrac{22}{23} = 0.96$

[1] These ratios fit the "normal" diagnosis row of Table x.8. Since this is known not to be the case, it can be seen that it fits closely, but not exactly, the "Partial Discharge - corona" diagnosis row.
[2] These ratios fit in two of the three categories in the "Low temperature thermal overheating" diagnosis row.
[3] These ratios fit the "High temperature thermal - more than 700°C " diagnosis in all three cases.

Table 3.24

Specifying and Testing New Oil

Avoiding the Pitfalls of New Oil Purchasing

In our business, when it comes to new oil, life is not quite as simple as we might hope it would be. It is usually not very wise to simply call up a transformer oil supplier and ask for delivery of some transformer oil. In the not too distant past, you could get away with that without expecting any serious difficulty. Now, the situation is more complex than that. If the purchase is significant, a simple approach is going to be less than ideal.

For several years, the supply side for transformer oil in the United States has been in an extended period of consolidation and redistribution. Major suppliers have left the industry, while new suppliers have entered. While demands for good quality transformer oil have largely been met, there have also been enough horror stories about errors and poor quality of oil purchased to raise considerable concern. Many of the problems that purchasers have experienced could have been either avoided or at least mitigated by properly specifying new oil purchases and performing adequate acceptance testing.

There are a number of possible sources for new oil specifications or acceptance standards. Using any or all of these to formulate actual purchasing specifications could be appropriate, depending on the purpose of the specifications and application of the oil. Industry consensus standards include ASTM Standard D 3487, Standard Specification for Mineral Insulating Oil Used in Electrical Apparatus, which includes specifications for Type I oil (0.08% by weight inhibitor, maximum – usually considered to be uninhibited) and for Type II oil (0.30% by weight inhibitor, maximum – considered to be inhibited even though there is no minimum level or acceptable range included). The ANSI/IEEE consensus standard for new oil is contained in C57.106 – 2002, IEEE Guide for Acceptance and Maintenance of Insulating Oil Equipment. The IEEE guide includes test limits for Type I and Type II oil from suppliers, as well as for new oil in new equipment and for new oil processed for filling into new equipment. IEC international consensus standard is IEC 60296, Edition 2, Fluids for Electrotechnical Applications – Unused Mineral Insulating Oils for Transformers and Switchgear. A revision of this (Edition 3) was distributed for voter approval as a final draft international standard in the latter part of 2003. As of press time, the outcome of that most recent attempt to revise the IEC standard is not known.

Industry consensus standards are an important consideration when formulating purchasing specifications, but they may not adequately represent the state-of-the-art when it comes to current supplies and suppliers. Industry experts can usually react more quickly to changing market conditions and both commercial and technical problems. For a number years, Doble Engineering Company has kept a current Transformer Oil Purchasing Specification (Doble TOPS). It has been widely used, particularly by large purchasers of new transformer oil, and the most up-to-date version is accessible on their website.

Similarly, S. D. Myers, Inc. maintains a current set of suggested new oil specifications. The current version as of press time is presented below. All of the industry consensus standards and the suggested specifications offered by industry experts have subtle, but important differences. Assuming that all potential problems with supply are addressed by inclusion of a reference to a specification usually turns out to be a bad decision. Oil characteristics, and their effects on performance in new or existing equipment, are of sufficient importance to justify putting together a reasoned and reasonable new oil purchasing specification appropriate to the buyer's application(s).

S. D. Myers, Inc. Suggested New Oil Purchasing Specification

The following five tables list the suggested specification values for the electrical properties, chemical properties, physical properties, and oxidation stability of new inhibited and uninhibited transformer oil. These, including the notes after the tables, represent the best information available to S. D. Myers, Inc. at press time. These specification values are for new oil, as received from the supplier. For processing new oil obtained using this purchasing specification for filling and for new oil in new equipment, the limits in IEEE Standard C57.106 – 2002 should be followed.

Electrical Properties of New Mineral Oil Dielectric Fluid[1]

Property	ASTM Standard Method	Specification Value
Dielectric Breakdown Voltage, disk electrodes, 0.10" gap	D877	30 kV minimum
Dielectric Breakdown Voltage, VDE electrodes, 1 mm gap	D1816	20 kV minimum
Dielectric Breakdown Voltage, VDE electrodes, 2 mm gap	D1816	35 kV minimum
Liquid Power Factor @ 25 °C	D924	0.05% maximum
Liquid Power Factor @ 100 °C	D924	0.30% maximum
Gassing Tendency	D2300	Negative[2]

Chemical Properties of New Mineral Oil Dielectric Fluid[1]

Property	ASTM Standard Method	Specification Value
Aniline Point	D611	84 °C max
Corrosive Sulfur	D1275	Passes[3]
Neutralization (Acid) Number	D974	0.015 mg KOH/g max
Moisture	D1533	25 ppm max
Polychlorinated Biphenyls (PCBs)	D4059	None Detected
Furanic Compounds, Total	D5837	20 parts per billion
Inhibitor Content (inhibited oil only)	D2668 or D4768	0.20 to 0.30% by weight[4]

Physical Properties of New Mineral Oil Dielectric Fluid[1]

Property	ASTM Standard Method	Specification Value
Color	D1500	0.5 maximum
Flash Point	D92	145 °C minimum
Interfacial Tension	D971	40 mN/m minimum
Pour Point	D97	-40 °C maximum
Kinematic Viscosity @ 40 °C	D445	11cSt maximum
Appearance	D1524	clear and bright

Oxidation Stability of New Uninhibited Mineral Oil Dielectric Fluid[5]

Property/Method Description	ASTM Standard Method	Specification Value
Sludge Free Life	Doble Procedure	40 ± 8 hrs minimum
Power Factor Value Oxidation	Doble Procedure	Passes[6]
Sludge @ 72 hours	D2440	0.15% maximum
Acid Number @ 72 hours	D2440	0.50 mg KOH/g maximum
Sludge @ 164 hours	D2440	0.30% maximum
Acid Number @ 164 hours	D2440	0.60 mg KOH/g maximum
Rotating Vessel	D2112	195 min. minimum

Oxidation Stability of New Inhibited Mineral Oil Dielectric Fluid[5]

Property/Method Description	ASTM Standard Method	Specification Value
Sludge Free Life	Doble Procedure	80 ± 8 hrs minimum
Power Factor Value Oxidation	Doble Procedure	Passes[6]
Sludge @ 72 hours	D2440	0.10% maximum
Acid Number @ 72 hours	D2440	0.30 mg KOH/g maximum
Sludge @ 164 hours	D2440	0.20% maximum
Acid Number @ 164 hours	D2440	0.40 mg KOH/g maximum
Rotating Vessel	D2112	220 min. minimum

Notes referenced in above tables:
1. Except for inhibitor content, the specification values for the electrical properties, chemical properties, and physical properties are the same for both uninhibited and inhibited oil.

2. The consensus standard from ASTM Standard D 3487 is for a positive evolution of gas of no more than 30 microlitres per minute from a 5 millilitre test specimen at 80 °C when using a 10 kV test voltage. This is the specification value for low, but still positive, gassing tendency under test conditions. It is highly desirable in large power equipment to have a negative gassing tendency oil – one that absorbs gas under conditions of the test rather than generating additional gases. Therefore, instead of using the consensus standard, the recommended specification for oil used in large power equipment is a negative gassing tendency oil.

   This specification value is for ASTM D 2300, Standard Test Method for Gassing of Electrical Insulating Liquids Under Electrical Stress and Ionization (Modified Pirelli Method). This should not be confused with total dissolved gas content – expressed as a percentage of oil volume – which is an acceptance value for oil processed for equipment 230 kV class and above prior to energizing from IEEE Standard C57.106, Guide for Acceptance and Maintenance of Insulating Oil Equipment.

   Similarly, Gassing Tendency as above should not be confused with Stray Gassing. In the 18-24 months prior to press time, there has been an increased interest in measuring the stray gassing tendency of new oil and appropriate methods of quantifying stray gassing have been developed. Stray gassing is the generation of gas because of low level thermal stresses on the oil. It becomes a problem if gases are generated early in the service life of the transformer by stray gassing. It can seem in such a case that there is a significant operational or design problem with new equipment where the true reason for the gases is a characteristic of the new oil.

3. ASTM D 1275, Standard Test Method for Corrosive Sulfur in Electrical Insulating Oil, is run at 140 °C. Because of difficulties experienced by equipment owners with corrosive sulfur developing in new oil, some laboratories are modifying the standard method to run it at 150 °C. The applicability of the modified method to the purchase specification needs to be negotiated between the buyer and vendor prior to any agreement concerning supply of new oil.

4. Inhibited oil should be in the range of minimum 0.2% to maximum 0.3% inhibitor. Uninhibited oil should be in the range of minimum "none detected" to maximum 0.08%. The maximum values do not apply if the base oil passes the appropriate oxidation stability tests with no higher inhibitor content than the levels indicated in these notes.

5. For purposes of evaluating the oxidation stability of uninhibited and inhibited transformer oil, it is very important that the oil adhere to the maximum levels of inhibitor – 0.08% by weight for uninhibited oil and 0.30% by weight for inhibited oil. There are separate specification values for oxidation stability of each type. Loading either type of oil with additional inhibitor may skew the oxidation stability results and may also cause an inferior oil to pass the specification limits for oxidation stability. Such an oil may provide poor service.

6. The procedure specifies maximum liquid power factor values for the oil after being subjected to controlled and accelerated oxidation. During the test, power factor values are plotted versus time and compared to a standard chart of acceptable values.

Notes not referenced by the tables:
7. Particle count distribution is typically not performed on new oil from suppliers. Rather, many transformer owners and laboratories are establishing limits for particle count distribution for oil newly installed in equipment both before energizing and at short intervals after energizing.

8. For new installations, manufacturers frequently have requirements for testing the oil in-service at short intervals after the unit is energized initially. In the absence of warranty requirements to the contrary, the minimum S. D. Myers, Inc. recommendation for new installations is that they be tested using dissolved gas analysis one month after being energized. This is in addition to any acceptance testing done immediately before or after the installation of the oil and initial energizing of the unit.

It is good practice, for large or critical installations, to keep a retain sample of the oil from the supplier and the oil that was processed for filling at least until after the one month sample for dissolved gas analysis is performed. A one quart glass bottle or the equivalent in smaller containers is appropriate as a retain sample.

9. If the one month DGA shows excessive gassing, part of the retain sample should be used to evaluate the stray gassing potential of the oil. Archiving the retain sample beyond the one month DGA is a decision that needs to take into account any acceptance testing that was done on the new oil.

Acceptance Testing of New Oil and New Installations

How does the buyer know that the oil being purchased meets the specifications that were agreed upon? The only way to know for sure is to perform acceptance testing. Generally speaking, a higher level of acceptance testing, checking more properties, is preferred.

However, the testing program must also be geared toward reality – both regarding the expense of the acceptance testing and for the lead time required to obtain some of the test values. In the real world, nobody should agree to perform $2000 worth of analytical work in order to prove that six drums of transformer oil meet specification – the testing would cost more than the oil is worth. Similarly, nobody should require that tankers or processing crew remain idle at a job site while a test on the oil is being run if that test will take a week or more to yield results.

Change the situation somewhat, however. The same owner may elect to do all of that expensive testing if the unit involved is a multi-million dollar GSU. Another oil purchaser may require the week long test to qualify a new supplier for annual or multi-year purchases of transformer oil for an entire electrical system.

This is one area where the purchaser and supplier need to determine what is appropriate for a particular situation. For buying a couple of small distribution units that are already oil filled, perhaps the owner accepts an oil testing certificate from the transformer manufacturer. For making a major installation requiring the services of the equipment and oil manufacturers, oil transportation, an installation contractor, the owner's own personnel, and maybe one or more subcontractors, a complete and extensive oil testing program at each step of the installation is appropriate. In addition to giving greater assurance that the job is going to go well by monitoring everyone's efforts, such a program also has a more practical benefit. If something goes wrong, who pays to correct the problem? If the purchaser cannot identify where a problem occurred, there is no choice for the purchaser other than absorbing the cost of fixing it.

Case History – But No Names, Please

Once upon a time, a major utility in the Eastern part of the United States bought and installed four identical transformers at the same time, and at the same facility. To make the job of installation easier, the company placed temporary bulk storage at the site and had the oil delivered to the bulk storage by tank truck. Throughout the installation, everything went apparently well until the units were electrically tested prior to being energized. The insulation power factor for all four units was around 2%, well in excess of the acceptance values. The problem was immediately linked to the oil. Using the liquid cup for the instrument at the site yielded unacceptable power factor results, and a laboratory test confirmed that the 100 °C liquid power factor was extremely elevated, indicating contamination of the oil. After extensive oil testing in several laboratories, the problem was diagnosed as a heel of unleaded gasoline in one of the tankers. The problem was mitigated to a great extent by fullers earth treatment of the oil in all four transformers. Eventually, the units were put in service, and have apparently operated well even though they have always had high liquid and insulation power factors.

So what is the problem, since everything turned out okay? The program to do the additional testing to identify the problem and then perform the oil work to a sufficient degree to allow the owner to actually energize and use the new transformers cost well into six figures. These necessary tasks would cost more if done today. The truly sad part of this tale is that the whole thing could have been prevented, just by using the tools at hand. All during the time oil was being delivered, there was an insulation power factor test set, with a liquid cup, at the site. If the purchaser had done something as simple as run a sample from each tanker using the liquid power factor cup in the insulation power factor test set, the tanker with the gasoline contamination would have been found and rejected. The contaminated oil never would have made it into the system in the first place.

Tests to Run on New Oil from Suppliers

It is not intended that these be considered to be in any order of priority or importance.

Acceptance Testing for New Oil when Received from a Supplier
- Corrosive Sulfur, ASTM Method D 1275 (can be modified to 150 °C)
- Inhibitor Content, ASTM Method D 2668 or D 4768
- Moisture, ASTM Method D 1533
- Liquid Power Factor at 25 °C and 100 °C, ASTM Method D 924
- Interfacial Tension, ASTM Method D 971
- Dielectric Breakdown Voltage, Disk Electrodes, ASTM Method D 877
- Acid Number, ASTM Method D 974 or D 664, or Approximate Acidity, ASTM Method D 1534
- Color, Appearance, ASTM Methods D 1500/D 1524

Qualifying Tests for New Supplies or for Purchases for Large Units
- Aniline Point, ASTM Method D 611
- Kinematic Viscosity, ASTM Method D 445
- Gassing Tendency, ASTM Method D 2300
- One or more of the Oxidation Stability Tests:
    - Sludge Free Life, Doble Procedure
    - Power Factor Valued Oxidation (PFVO), Doble Procedure
    - Oxidation Stability, ASTM Method D 2440
    - Rotating Vessel Oxidation, ASTM Method D 2112

Optional Test or Tests to Diagnose Specific Problems with New Oil
- Flash or Fire Point, ASTM Method D 92
- Dissolved Copper, ASTM Method D 3635
- Dissolved Metals by ICP, ASTM Method Pending
- Particle Count Distribution, ASTM Method D 6786
- Inorganic Chlorides and Sulfates, ASTM Method D 878
- Carbon Type Distribution, ASTM Method D 2140

- Particles and Filming Compounds Analysis, Proprietary Method
- Stray Gassing, Proprietary Method

Testing Load Tap Changer Dielectric Liquid

The traditional testing program for load tap changer oil has been to perform: the basic chemical, physical and electrical tests that comprise our Oil Screen (OS) package, a moisture determination by Karl Fischer analysis, and a dissolved gas analysis. The importance of these tests, performed on a regular basis, cannot be minimized. However, after a substantial development and evaluation process, we have made an addition to our routine recommendation for testing dielectric liquids from load tap changers. For convenience, we have created an LTC Testing Package that includes both the traditional tests and the additional new testing protocols. After discussing the purpose and significance of the new testing package, we will review the same information for the traditional approach.

New Testing Protocols

Our recommended addition to customers' current periodic testing program is a package of additional oil tests and analytical services. These were developed specifically for diagnosing the need for LTC maintenance. There are essentially three tests in the new package.

First, a particle count and distribution is measured for the oil sample. The reason that particle count is important for LTCs in particular is that certain types of wear and also oil oxidation and breakdown will create microscopic particles that are suspended in the oil. Knowing how many are present and size distribution of the particles is useful in interpreting what is going on inside the equipment and whether unusual conditions or operational problems are responsible for the particles that are being created.

Normal operations yield a certain distribution as to size and number. Abnormalities in the operation affect this profile. We can interpret the particle count distribution and use it with the other test data to diagnose specific problems and get an indication of the rate of wear.

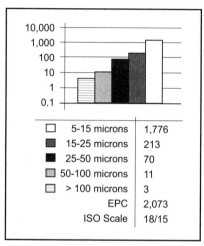

Figure 3.29 - Particle count distribution for a typical LTC, in table and graph formats.

Second, individual particles are identified and examined microscopically. A combination of techniques are used to separate particles found suspended in the oil so that a scanning technique can be used to identify, characterize, and document the kinds of particles present.

Particles are characterized with regard to their composition – metallic, non-metallic, carbon; ferrous, non-ferrous metals; created by wear or arcing; etc. The wear pattern can be determined, and this will indicate whether the wear is expected or not. Such an examination will also indicate whether the arc is being quenched properly. There are a number of different conditions to look for that this testing will indicate.

NOTE: Both of the conditions shown in the example above are abnormal for load tap changer operations. The example of

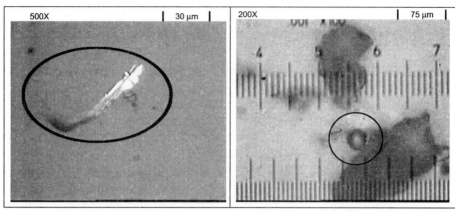

Figure 3.30 - Example of a ferrous, shearing wear particle (left). Examples of iron and copper "techtites" generated from electrical arcing (right).

arcing products includes iron spheres formed during arcing. This indicates that either the arc is not being quenched soon enough or that the contacts are completely worn so that the arc formed at breaking the circuit is involving the contact supports.

The example of ferrous shearing wear particle indicates a mechanical malfunction of the tap changer operation.

Third, the oil is subjected to an additional chemical analysis that will indicate its tendency to form films or insoluble compounds that are electrically and mechanically resistant. This has an application to the oxidation of the oil and adds information about how well the oil is capable of performing its function in the LTC.

The test for coking and filming indicators is important. Values for Absorbance that are above 1, and any increases in Absorbance over time, indicate that the oil has chemically changed while in service and has the potential to form an electrically and/or mechanically resistant film on contacts. This condition requires more mechanical force to move contacts and results in poor arc quenching, both of which lead to increased wear on contacts and overheating.

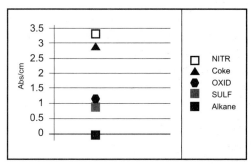

Figure 3.21 - Analysis for coking and filming indicators.

Taken together, the results of these tests and the interpretation and diagnosis of conditions in the load tap changer indicate whether the regularly scheduled maintenance interval for the oil and for the mechanical and electrical components of the LTC is appropriate or whether closer monitoring or a shortened maintenance interval is required. As is typically the case with most testing of dielectric fluids, the values for a given set of tests are very important, but changes in these values over time are even more useful for diagnosing the conditions that affect load tap changer operations.

Traditional Oil Testing Methods

The new testing protocols indicate a number of conditions that the traditional methods of analysis do not address. However, the traditional oil testing methods are still very useful and should not be ignored, since they also indicate and diagnose some conditions that the newer methods are not designed to measure.

The liquid screening tests of the Oil Screen package include the neutralization number (acid number), interfacial tension (IFT), D 877 dielectric breakdown strength, specific gravity (relative density), color, and appearance. These tests measure the progress of oil oxidation (particularly the acid number and IFT) and indicate the presence of contamination from outside sources (particularly the D 877 value and specific gravity).

The moisture analysis by Karl Fischer titration is a direct measure of the moisture content of the dielectric liquid used in the load tap changer. At high levels of saturation, moisture affects the ability of the oil to provide adequate dielectric strength and to efficiently extinguish the arcs that form as the LTC operates. At lower levels of saturation, an increase in moisture content contributes to a more rapid ageing and oxidation of the oil.

Dissolved gas analysis continues to be useful as a diagnostic tool for load tap changer operations and maintenance. Normal dissolved gas profiles have been established for many different manufacturers and models of LTCs. Additionally, diagnosis of abnormalities in operations based on changes in the gas profiles is reasonably well understood. A number of standards and technical organizations (including IEEE) continue to address these issues, and S. D. Myers, Inc. is a significant participant in and contributor to these efforts. The main emphasis of dissolved gas analysis of LTC dielectric liquids is to identify abnormal overheating of the contacts and the oil, and dissolved gas analysis is very effective at doing this. Combining dissolved gas analysis – which indicate the severity of the abnormal condition – with the newer testing protocols – which indicate the probable cause of the abnormal condition and may indicate the severity – is particularly useful.

Additional testing, including liquid power factor determination, oxidation inhibitor content, and metals analysis may be recommended to address specific system needs concerning maintenance of LTCs and LTC insulating liquids.

Oil Test Methods and Standards

Description	ASTM Number	New Oil	In Service	Comments
Aniline Point	D611	physical property, composition, function	no	Aniline point defines wherther an oil is naphthenic and describes temperature to use for hot oil cleaning (dissolve sludge). This is a specification test for new oil.
Color	D1500 D1524	physical property purity	routine	New highly refined oil is very low color. As oil ages, it darkens. Part of Oil Screen package, and described in In-Service Testing section.
Visual Examination	D1524	pass/fail	routine	Anything other than clear is not acceptable. Part of Oil Screen package, and described in In-Service Testing section.
Turbidity	D6181	--	--	Relatively new method (1997), not widely used as yet. Most use D1524 (Visual Examination) instead. D6181 intended to measure changes not visible to the naked eye.
Sediments and Soluble Sludge	D1698	no	--	Determines approximate inorganic and organic composition of sediment and soluble sludge. Not used much as an in-service test any more.
Particle Count Using Automatic Optical Particle Counters	D6786	perhaps	not routine, diagnostic	The optical method and its interpretation are much better developed than the ESZ method. Used for trouble shooting transformer oil. Also may be a routine in-service test for other oil filled equipment.
Particle Count (ESZ - Electrical Sensing Zone)	Proposed	--	--	The application of this method not the same as the optical method. This is a new application of this technology to insulating oil and is not yet widely used.
Particle characterization and filming compound analysis	In-house methods, so far.	troubleshooting	not routine, diagnostic	Describes the types of particles present and evaluates the tendency of the oil to for resistive films. Used for troubleshooting problems in transformer. May be a routine test for other oil filled equipment.
Dissolved Metals by ICP	Proposed	perhaps	periodic, diagnostic	ICP technique is well established for detecting a number of metals in oil. Used as a diagnostic test and described in In-Service Testing section.

Oil Test Methods and Standards

Description	ASTM Number	New Oil	In Service	Comments
Dissolved Copper by Atomic Absorption	D3635	perhaps	periodic, diagnostic	Approved method is for copper, only. Some have adapted to other metals. Similar to ICP.
Corrosive Sulfur	D1275	chemical property, composition, purity	rarely	Often, both a specification value and an acceptance test for new oil in United States.
Dielectric Breakdown Using Disk Electrodes	D877	electrical property, function, purity	routine	D877 still is useful for testing in-service oil. Not a good test for water or oxidation products. Optional acceptance test. Part of Oil Screen package, and described in In-Service Testing section.
Dielectric Breakdown Using VDE Electrodes	D1816	electrical property, function, purity	routine	More sensitive than flat disk method. Sensitive to dissolved gases - not a cure all. Acceptance test and important in-service test. Described in In-Service Testing section.
Dissipation Factor (Power Factor)	D924	electrical property, function, purity	routine	Important acceptance and in-service test at both temperatures - 25° C and 100° C. Described in In-Service Testing section.
Relative Permittivity (Dielectric Constant)	D924	electrical property, function	no	Design consideration.
Specific Resistance (Resistivity)	D1169	electrical property, function	rarely	DC test. Liquid power factor is used more often in U. S. for acceptance and in-service testing.
Dissolved Gas Analysis (DGA)	D3612	as gas content	routine, diagnostic	DGA is an important in-service oil test for both routine monitoring and for diagnosis of fault conditions. DGA is used as a substitute for some new oil gas content tests such as D831, D1827, and D2945. Described in section on Dissolved Gas Analysis
Flash Point and Fire Point by Cleveland Open Cup	D92	physical property, composition, function	rarely	Flash point and fire point are frequently specification values. Flash point is an optional acceptance test.
Furanic compounds	D5837	optional	periodic, diagnostic	Important in-service oil test for monitoring aging and degradation of solid insulation and for diagnosing fault conditions involving the paper. Described in the section on furanic compounds analysis.
Gassing Tendency	D2300	electrical property, composition	no	Frequently a specification value. New oil test used for troubleshooting

Oil Testing | 173

Oil Test Methods and Standards

Description	ASTM Number	New Oil	In Service	Comments
Stability Under Electrical Stress	D6180	--	--	New method (1998), not yet widely used.
Stray Gassing	Pending	--	--	Troubleshooting test for new oil.
Oxidation Inhibitor by Gas Chromatography	D4768	chemical property, composition	routine	May be used interchangeably with the Infrared Absorption method to quantify inhibitor. Described in the in-service testing section.
Oxidation Inhibitor by Infrared Absorption	D2668	chemical property, composition	routine	May be used interchangeably with the Gas Chromatography method to quantify inhibitor. Described in the in-service testing section.
Inorganic Chlorides and Sulfates	D878	chemical property, composition	rarely	Sometimes used as a specification value, more often used to troubleshoot specific conditions.
Interfacial Tension (IFT) by the Ring Method	D971	physical property, purity	routine	Very useful specification value, acceptance test, and routine monitoring test. Part of Oil Screen package, and described in In-Service Testing section.
IFT by Water Drop Method	D2285	physical property, purity	routine	May be used interchangeably with the Ring Method.
Kinematic Viscosity	D445	physical property, function	no	Specification value related to heat transfer. D445 Kinematic Viscosity is the preferred viscosity measurement due to accuracy.
Saybolt Viscosity	D88	physical property, function	no	Can be a substitute for Kinematic Viscosity, but this is not the preferred solution.
Conversion of Kinematic Viscosity to Saybolt Viscosity	D2161	--	--	Practice for converting viscosity values.
Neutralization Number	D974	chemical property, purity	routine	Very useful specification value, acceptance test, and routine monitoring test. Part of Oil Screen package, and described in In-Service Testing section.
Acid number by Potentiometric Titration	D664	chemical property, purity	routine	Equivalent to and interchangeable with D974.
Approximate Acidity	D1534	no	screening test	Can be useful as a preliminary screening device to define units that need more in-depth testing.
Oxidation Stability	D2440	chemical property, stability	no	Specification value related to oil life and stability.

Oil Test Methods and Standards

Description	ASTM Number	New Oil	In Service	Comments
Oxidation Stability by Pressure Vessel	D2112	chemical property, stability	no	Specification value related to oil life and stability.
Sludge Free Life	Doble Procedure	chemical property, stability	no	Specification value related to oil life and stability.
Power Factor Valued Oxidation	Doble Procedure	chemical property, stability	no	Specification value related to oil life and stability.
Polychlorinated Biphenyls (PCBs)	D4059	chemical property, regulatory compliance	periodic or for regulatory compliance	Prefered method for PCBs in electrical equipment dielectric fluids.
Pour Point	D97	physical property, function	no	Specification value related to heat transfer and low temperature performance.
Refractive Index and Refractive Dispersion	D1218	physical property, composition	no	Used primarily for R&D and to troubleshoot.
Refractive Index and Specific Optical Dispersion	D1807	physical property, composition	no	Used primarily for R&D and to troubleshoot.
Carbon-Type Distribution	D2140	physical property, composition	no	Used primarily for R&D and to troubleshoot.
Relative Density (Specific Gravity)	D1298	physical property, composition, function	routine	Specification value and acceptance test related to heat transfer. As a monitoring test, sensitive to outside contamination. Part of Oil Screen package, and described in In-Service Testing section.
Sampling Electrical Insulating Liquids	D923	--	--	Practice for sampling liquids for general purpose testing.
Sampling Electrical Insulating Liquids for Gas Analysis and Determination of Water Content	D3613	--	--	Practice for sampling liquids for testing by methods sensitive to atmosphere contamination such as DGA and moisture content.
Specification for Mineral Oil	D3847	--	--	Standard specification.
Terminology	D2864	--	--	Definition of terms used in ASTM standards.
Water by Karl Fischer Titration	D1533	chemical property, purity	routine	Very useful specification value, acceptance test, and routine monitoring test. Described in In-Service Testing section.

Oil Testing | 175

Chapter 4

Electrical Testing of Transformer Insulation

Part 1 – Why Perform Electrical Tests?

Introduction

Power transformers are usually very reliable pieces of equipment. However, as the transformer ages and approaches its expected end of life, its components deteriorate and the likelihood of failure increases. The main problems are cellulose deterioration and core and winding movement. Other problems, like worn tap changer parts and loose connections, bushing problems and high moisture content can also be a source of failure.

Periodic field electrical testing of transformers monitors conditions in the insulation and evaluates the useful life available in the equipment tested. Whether or not the useful life, as revealed by the tests, guarantees against future operating failure, depends upon sound engineering judgement. That is why the plant engineer and testing technician from a consulting/testing company – drawing on their collective experience and working with the equipment's specifications – ultimately must determine the condition of the transformer.

So, if you've already run oil tests on the equipment while it was energized and received a clean bill of health, your transformers may still not be in optimal condition. The following brief review of oil tests will reveal their shortcomings.

1. Oil Screen Testing

Good oil screen test results (acid, IFT, dielectric, relative density, color, visual, and sediment) do not necessarily signify good, dry windings in your transformer. You also need to find out if the cellulose contains moisture. Has the insulation shrunk? Is it brittle? Does it have holes in it? Oil screen testing in short, cannot convey the whole story regarding cellulose insulation.

2. Moisture-in-Oil (Karl Fischer) (ASTM D 1533-96)

This test may show that the transformer oil is dry, but what about the solid insulation? Most people wouldn't worry about a moisture content of 25 ppm. However, moisture saturation of the oil is temperature dependent. This means that at a temperature of 60 °C (when the oil can hold a lot of water), 25 ppm wouldn't be a problem. The percent moisture by dry weight (%M/DW) would only be 0.62. But 25 ppm in the oil at 20 °C (when the oil cannot hold a lot of water) would be a problem. The %M/DW would be 5.37. If the unit has 10,000 lbs of insulation, the difference between 0.62% (7.4 gallons of water) and 5.37% (64 gallons of water) is 57 gallons of water locked in the solid insulation!

3. Gas-in-Oil Analysis (ASTM D 3612-96)

Even though dissolved gas analysis (DGA) has proven reliable for detecting incipient faults, insidious (gradual and cumulative) ones may not be detected even with continuous on-line gas monitoring. When the rate of oil and cellulose deterioration insufficiently produces the heat and corresponding gases associated with DGA, electrical testing may expose a problem.

4. Thermographic Survey (Infrared)

The infrared (IR) inspection of in-service equipment has proven to be a valuable tool in the substation maintenance program. IR survey of the transformer and its associated accessories, however, may not reveal every problem. IR is just an external view of the transformer; what's happening internally may be hidden.

Other reasons for performing field electrical tests are as follows:

- Acceptance test
- Prior and subsequent to physical movement
- Periodically monitor insulation changes
- Transformer drops off-line for any reason
- Prior and subsequent to prolonged storage
- Smell/sound of arcing
- Evidence of excessive moisture (free water, cloudy oil)
- Prior and subsequent to prolonged or excessive overheating/overloading
- Any unsatisfactory oil test data (DGA, H_2O, color change)

A Complete Physical

When you want a complete bill of health, you undergo a thorough physical; likewise with your transformer. Since no single test procedure adequately supplies all the information needed to properly evaluate a transformer, a basic battery of tests should be conducted.

The frequency of these tests will be determined by many factors, such as the type and class of equipment, its importance, age, load, and history of operation. Scheduled outages should be utilized as periods for electrical testing. If a fault appears to be developing, more frequent electrical testing may be scheduled to monitor this condition.

Part 2 - A General Overview

Records

Records of the test results shall include the following:

- Scope of the testing
- Equipment tested and date of test
- Test results
- Interpretation and recommendations based on the testing

This information should be permanently recorded for each of the transformers. Records should include all factory tests as well as factory drawings of the transformers.

Safety

General Precautions

Some tests, like DGA, Oil Screen, and Infrared Thermography do not require that the unit be de-energized. This is not the case for electrical testing. Special attention and training is needed to ensure protection. Personnel performing such testing must observe proper safety regulations at all times. Plant safety rules as well as state and federal regulations should be consulted when working on and around energized equipment. Temporary safety barriers may be needed to keep other personnel a safe distance from the test area.

The testing devices in use today are highly sophisticated. Restrict the use of this equipment to properly trained personnel.

The test voltages generated by this equipment represent a hazard to both the equipment under test and testing personnel. Equipment to be tested must be disconnected from the power system using established operating procedures. Proper grounding techniques must be practiced to avoid injury to personnel or damage to equipment. Personnel should be trained to treat all ungrounded equipment as energized. Never take the word of another that something is grounded or de-energized. Prove that it is for yourself.

Lockout/tagout procedures shall be followed as required by plant, state, and federal safety procedures. Precautions must be taken to prevent personnel from contacting any terminals of the apparatus under test. When possible, an observer should be stationed to warn approaching personnel of the potential hazard. This person should be provided with a means of de-energizing the power source.

Entering Transformers

New transformers are sometimes shipped without oil to reduce the shipping weight. To prevent the entrance of moisture during transit, the tank may be filled with dry nitrogen gas under low pressure, displacing oxygen. On the other hand, when oil is drained from service-aged transformers prior to making an internal inspection, carbon monoxide and/or other combustible gasses may evolve from the cellulose in concentrations that are also dangerous. Because of such atmospheric conditions and other potential hazards, transformers must generally be classified as permit entry confined spaces and treated accordingly under safety regulations prior to entry.

Objectives

Electrical tests on transformers will fulfill four distinct but general functions:

- Prove the integrity of the piece of equipment at the time of acceptance
- Verify the continued integrity of the unit at periodic intervals of time and plot any changes

- Determine the nature or extent of the damage when a unit has failed
- Validate a successful field repair

The specific tests required to fulfill each function may vary and will be tailored to meet the equipment's requirements and the technician's recommendations.

Acceptance Tests (Factory and Field)

We recommend running field acceptance tests on a transformer immediately upon arrival and prior to removing from the carrier. Damage may occur during transit that may be detected during field acceptance testing. Test data should be compared to the factory acceptance testing data. If a change is apparent, contact the manufacturer.

Manufacturers are capable of producing a properly dried transformer with a moisture content of 0.5% M/DW or less. However, experience also tells us they are capable of producing a unit in excess of the 0.5% M/DW limit (Figure 4.1). We are responsible as consumers for indicating in our purchase specifications a moisture content of 0.5% M/DW or less. Running a power factor or dissipation factor test is a good way of verifying the unit's moisture content. If field acceptance testing reveals a power factor in excess of 0.5%, you should take steps to dry the unit properly.

We recommend that transformer purchase specifications stipulate that manufacturers' power factor testing method conforms to the purchasers' field-testing practice.

Periodic Tests

All insulation inevitably ages over time. Aging rates vary and are influenced by oxygen, water, heat, and the by-products of the oxidation process. The rate of aging is never static. Sometimes it's

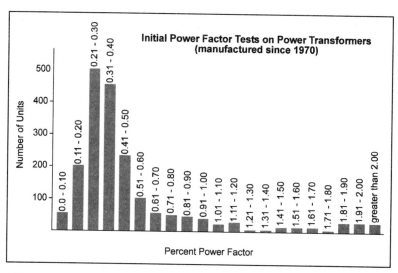

Figure 4.1 - Initial power-factor tests on power transformers manufactured since 1970 (courtesy of Doble Engineering Co.).

faster, sometimes slower, but it's always present. This is true of all transformers, whether energized or not. It is important to monitor the condition of the cellulose. Its condition will also indicate remaining life and signal when maintenance is needed.

Testing Units After Failure

Caution – When a transformer has gone off line, extreme care needs to be taken to determine the cause of the outage. Premature re-energizing may result in additional damage or complete failure. The possibility of serious injury to personnel also exists. Technical publications by NETA, IEEE, and NFPA address many troubleshooting problems. The manufacturer's assistance may also be procured.

Causes – These outages may be in response to primary breaker or fuse operation, pressure relay or mechanical relief operation, differential relay operation, etc., and may have been preceded by a voltage line problem. The inspection is intended to expose clues as to what triggered the operation of the protective device. This cause may range from failure of the transformer to a faulty protective device.

Clues - The initial visual inspection of the transformer's exterior may reveal such things as tank wall bulge, cracked or leaking seals, oil spills. Visual inspection through the manhole may reveal burned cellulose, melted metals, darkened oil, and unusual odor. Be cautious as combustible gases may be present in the oil and/or in the head space.

Appropriate electrical tests, to be detailed later, include transformer turns ratio, insulation resistance, winding resistance, and power factor. High voltage tests should be avoided as they may excessively stress an already weakened insulation. Instead, perform a dissolved gas-in-oil analysis and furan testing.

If determined that the outage was not due to a transformer fault, it may be re-energized safely.

Standards

Transformers that have been built according to American National Standard Institute (ANSI)/The Institute of Electrical and Electronic Engineers Inc. (IEEE) comply with the standards in the "IEEE Standards Collection for Distribution, Power and Regulating Transformers." This collection is also known as the C57 Series. Every attempt should be made to consult the latest version of the standards.

Application

The test requirement for any given repaired transformer may be unique. This is understandable, considering the complex variety of failure modes and repairs that may be involved. Consider the options available when a transformer fails.

- Repair in kind and risk the possibility of a repeat failure.
- Repair employing updated materials and design philosophy, which may void original performance guarantees.
- Scrap the unit and purchase a new one.

If you choose either of the first two options, you need to modify the acceptance tests to accommodate the new specifications.

Part 3 – Electrical Tests

Guidelines

When applicable, manufacturers' recommendations supplied in the equipment's instructions must be factored into testing procedures. The following guidelines are general in nature and must be modified to match specific equipment. The list below includes the most common tests used in the field to evaluate the condition of the transformer insulation, mechanical integrity, and ground systems.

AC Tests

- Insulation Power Factor/Dissipation Factor Test
- Power Factor Tip-up Test
- Single-Phase Excitation Current Test
- Transformer Turns Ratio Test
- Bushing Power Factor Test
- Liquid Insulation Power Factor Test

DC Tests

- Insulation Resistance Tests (Megger®)
- Dielectric Absorption
- Polarization Index (PI)
- Step Voltage Test
- Core Ground Test
- Overpotential Test (Hi-Pot)
- Winding Resistance

When coordinated with data from oil tests, a reliable picture of the transformer's internal condition begins to evolve. With this information, you decide on the services needed to correct the problems and when to schedule future tests.

Methods such as Frequency Response Analysis (FRA), Leakage Reactance, and Recovery Voltage Method (RVM) are beginning to impact the testing industry.

Insulation Power Factor Series

Insulation power factor should not be confused with system power factor in an AC network. Insulation power factor indicates the quality of the insulation and may be understood from the following explanation. Any winding in a transformer is separated from all other windings and ground potential by solid insulation. Cellulose insulation forms an effective capacitance network as indicated in Figure 4.2. All electrical insulation has a measurable amount of dielectric loss regardless of age.

The power factor tests measure the insulation capacitance, AC dielectric losses and ratio of the measured quantities. The power

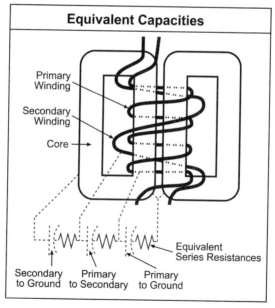

Figure 4.2 - Dielectric loss of each capacitor divided by capacitive voltamperes is equal to power factor.

factor is the cosine of the phase angle (cosine θ). You can measure it by first applying a voltage across this capacitance and measure the amperes and watts loss and calculate the power factor.

$$\frac{W}{VA} \times 100 = \% \, PF$$

This measurable dielectric loss will develop heat in the insulation during transformer operation. Heat, along with moisture and other factors can cause deterioration of the insulation. Interpreting the insulation condition depends primarily on comparing previous test results from the same unit or from similar units in good condition.

The de-energized series of power factor tests, often referred to as "Doble Tests," use one of various types of equipment available or can be calculated from readings taken with instruments that measure volts, current, and watts loss.

Another method called dissipation factor may also be used to measure the AC dielectric loss. It is also referred to as the tan delta (Tangent of the dielectric loss angle δ). The dissipation factor is very similar to the power factor tests. The two test methods are comparable up to 10%.

Winding Insulation Power Factor Test

As previously mentioned, this equipment measures the power loss through the insulating system to ground that is caused by dielectric loss or leakage current. All insulating systems will have minor leakage paths through the system that will permit a small current flow. If the insulating system is perfect (a theoretical condition), there would be zero current leakage. Then a power factor of 0% could be calculated. The greater the leakage path, the greater the power loss through the insulating system. This is caused by the I^2R watts loss as determined by Ohm's law. If all of the input power were lost, the power factor would be 100%.

A power factor reading of 0.3% is not unusual under ideal conditions. Data shown in Figure 4.4, line 6 gives a reading of 0.55%. Although not critical, the insulation on the secondary of this unit may be starting to deteriorate.

Tabulated in Figure 4.5 bottom are the standard series of tests (1 through 4), two calculated values and two supplementary tests. This illustrates how data pertaining directly to C_H, C_L, and C_{HL} is obtained.

Figure 4.3 - M4000 Power Factor Test Set (courtesy of Doble Engineering Co.)

Obviously, a condition of a high power factor existing in one component (C_H for example), may be masked in test 1 of Figure 4.5 but accentuated in test 2 of Figure 4.5. This phenomena permits separate determination of C_H, C_L, and C_{HL}. The obvious solution is to standardize the method illustrated in Figure 4.5. It is consistent with the Std. C57.12.90-1993 "IEEE Standard Test Code for Liquid-Immersed Distribution, Power and Regulating Transformers," Method II, Test with Guard Circuit. This is the test method used most commonly in the field. It increases the sensitivity of the power factor test to localized conditions. The method permits individual measurements of charging currents and dielectric loss for C_H and C_L. Power factors can be calculated from these values. Currents and losses for tests 2 and 6 are subtracted from those recorded for tests 1 and 5 to obtain data for a power factor calculation for C_{HL}. If desired, C_{HL} may be measured directly by the Ungrounded Specimen Test (UST) method. Figure 4.4 shows typical test data and calculated power factor test results. From the data in Figure 4.4, you can see that the measured value of C_{HL} by the UST method closely agrees with the calculated value in line 4. If the measured and calculated values agree, the test is performing properly.

Figure 4.6 illustrates a system for lumped measurements on a two-winding transformer. Three measurements are made: the first two (A and B) include ground and interwinding insulation and the

Insulation Power Factor

#	Test Connections			Test kV	Equivalent 10kV		% PF		correction factor	windings measured capacitance (pF)	measured	insulation rating
	energized	ground	guard	UST		mA	Watts	measured	corrected				
1	High	Low			10	41.0	2.0			0.80	10850	$C_H + C_{HL}$	
2	High		Low		10	15.8	0.8	0.51	0.41	0.80	4204	C_H	G
3	High			Low	10	25.0	1.1	0.44	0.35	0.80	6650	C_{HL} (UST)	G
4	(Test 1) - (Test 2)					25.2	1.2	0.48	0.38	0.80	6646	C_{HL}	G
5	Low	High			10	69.0	4.2			0.80	18240	$C_L + C_{HL}$	
6	Low		High		10	44.0	3.0	0.68	0.55	0.80	11600	C_L	D
7	Low			High	10	25.0	1.2	0.48	0.38	0.80	6660	C_{HL} (UST)	G
8	(Test 5) - (Test 6)					25.0	1.2	0.48	0.38	0.80	6640	C_{HL}	G
9	C_H - Bushing C1 measured					12.02	0.62	0.52	0.42	0.80	3201.2	C_H	G
10	C_L - Bushing C1 measured					40.08	2.36	0.59	0.47	0.80	10557	C_L	G

Figure 4.4 – Insulation power factor test data

third (C) includes the combined ground insulations of the high and low voltage windings. The power factors recorded for the three measurements are the weighted average of the two components involved in each.

New transformers should have power factors of 0.5% or less. Modern oil-filled transformers that have been in-service can have power factors as high as one percent and still be deemed worthy of continued operation. However, insulation power factors above one percent are cause for concern. Additional tests should be conducted to further investigate the probable cause of the power factor recorded. The bushings and liquid insulation should be tested to see if they contribute to the overall high power factor.

The effects of temperature on power factor measurements is considerably important. The magnitude of the power factor recorded on a given specimen varies directly with temperature. Temperature, therefore, must be taken into account when comparing data recorded for the same unit in the field with data from the factory.

It is standard practice to correct the measured power factor values to a value based on 20 °C. When a transformer is tested near freezing temperatures, the large correction factor may cause

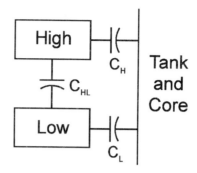

Test	Energized	Ground	Guard	Measure
1	H	L	-	$C_H + C_{HL}$
2	H	-	L	C_H
3	L	H	-	$C_L + C_{HL}$
4	L	-	H	C_L
5	Test 1 - Test 2			C_{HL}
6	Test 3 - Test 4			C_{LH}

Figure 4.5 - A simplified diagram of a typical two-winding transformer (top); C_H refers to all insulation between the high voltage winding and grounded parts, including bushings, windings, insulation, structural insulating members, and oil (bottom). C_L refers to the same parts and materials between the low voltage windings and grounded parts. C_{HL} refers to all winding insulation, barriers and oil between high and low voltage windings. (courtesy of Doble Engineering Co.)

Test	Energize	Ground	Measure
A	H	L	$C_H + C_{HL}$
B	L	H	$C_L + C_{HL}$
C	HL	-	$C_H + C_L$

Figure 4.6 – Alternative method for power factor testing; lumped measurements- more applicable for factory testing.

the power factor to be unreliable. In this case the test should be repeated at a higher temperature. Likewise, if the transformer is above 50 °C, the power factor could also be in error. This test should be repeated after the unit cools down.

There are temperature correction charts published for the various types of transformers and insulating fluids. While the problem of temperature correction does exist, it is not complicated. Refer to the test equipment manufacture for the most recent temperature correction curves.

The power factors recorded for routine overall tests provide interesting and valuable information regarding the general condition of ground and interwinding insulation of transformers. These values also provide a valuable index of dryness for a transformer. They are helpful in detecting undesirable operating conditions and failure hazards resulting from moisture, carbonization of insulation, defective bushings, contamination of oil by dissolved materials or conducting particles, improperly grounded or ungrounded cores, etc.

In conjuction with each power factor measurement, an evaluation of the insulation capacitance should also be made. The capacitance can be either directly measured during the power factor measurement or it can be calculated for the milliamperes reading using the formula (milliamperes x 265 = capacitance in picofarades). A capacitance analysis is based on a comparison with either factory test values or the initial field test. A capacitance change greater then 10% is indicative of winding deformation or movement.

In an effort to improve the sensitivity of the overall tests by minimizing the amount of insulation included in the measurement, particularly in larger transformers, it is essential that separate tests also be performed on the bushings and liquid insulation in a transformer.

Insulation Power Factor Tip-Up Test

On many occasions, insulation power factor tests for a transformer are higher than normal. This indicates that something is wrong in the insulating system. A short, easy to perform test called the insulation power factor tip-up test can provide more data that may help pinpoint the trouble.

Using the power factor test set, you can perform the tip-up test by varying the applied test voltage from zero volts to the maximum in several equal steps. Calculate the percent power factor at each voltage. If the power factor does not vary as the test voltage is changed, moisture or other polar compounds are the cause of the high power factor. If the power factor increases as the test voltage increases, then ionization is occurring, carbonization of the oil or windings, and/or voids or air gaps in the winding insulation of dry-type transformers.

Figure 4.7 shows test data for a transformer when ionization was occurring. You can see that the power factor values increased as the test voltage increased.

When the transformer was untanked at a service shop, burned areas and shorted turns were found and fine carbon residue was deposited throughout the winding insulation.

Test (kV)	Power (C_H)	Factors (C_T^1)	Corrected (C_{HT})
10	0.60	4.52	0.81
2	0.40	3.05	0.55
1	0.17	2.56	0.45
C_T^1 - Tertiary Winding			

Figure 4.7- Insulation power factor tip-up test results.

Bushing Power Factor Tests

Bushings can be tested in several different ways without removing them from the transformer. The most effective test method utilizes the capacitance taps or power factor taps. This method as shown in Figure 4.8 is referred to as the Ungrounded Specimen Test (UST).

Figure 4.9 lists typical field test data. Be particularly careful not to exceed the allowable tap test voltage. Consult the manufacturer's published information for the maximum permissible test voltage for the particular type of bushing being tested.

The hot-collar test was designed to detect localized defects (contamination or voids) in compound-filled bushings or cable potheads not equipped with power factor or capacitance taps. This test is applied, using single or multiple collars on transformer bushings (Figure 4.10), if the UST method cannot be performed. It is also used to supplement the UST test when compound-filled bushings are involved.

Since relatively low dielectric losses and currents are normally recorded for hot-collar tests, small changes in either value can result in misleading changes in calculated power factors. Because of this, we recommend that hot-collar tests be evaluated by comparing currents and losses obtained from similar tests on similar bushings and potheads under the same atmospheric conditions. Power factor values need not be calculated. As a general

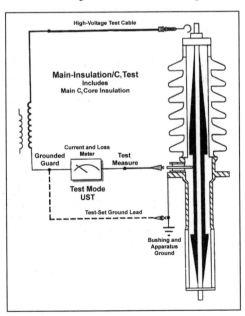

Figure 4.8 – Ungrounded specimen test on transformer bushings. (courtesy of Doble Engineering Co.).

Bushing Tests

		Nameplate		Test kV	Equivalent 10kV		% Power Factor		correction factor	measured capacitance (pF)	insulation rating
		capacitance (pF)	% power factor		mA	Watts	measured	corrected			
H_1	C1	332	.52	10	1.260	0.060	0.48	0.47	0.98	332.00	G
H_2	C1	331	.52	10	1.260	0.060	0.48	0.47	0.98	334.20	G
H_3	C1	333	.52	10	1.260	0.060	0.48	0.47	0.98	336.60	G
H_0	C1										
X_1	C1			10	1.300	0.140	1.08	0.99	0.92	346.60	G
X_2	C1			10	1.280	0.140	1.09	1.00	0.92	340.00	G
X_3	C1			10	1.340	0.160	1.19	1.09	0.92	356.40	G
X_0	C1										
H_1	C1	1240		0.5	4.800	0.080	0.17	0.17	1.00	1296.0	G
H_2	C1	1100		0.5	4.000	0.080	0.20	0.20	1.00	1122.0	G
H_3	C1	1010		0.5	4.000	0.080	0.20	0.20	1.00	1055.0	G
H_0	C1										
X_1	C1			0.5	8.000	0.800	1.00	1.00	1.00	2067.0	G
X_2	C1			0.5	8.000	0.880	1.10	1.10	1.00	2136.0	G
X_3	C1			0.5	8.000	0.880	1.10	1.10	1.00	2140.0	G
X_0	C1										

Figure 4.9 - Typical field test data for a large transformer bushing. Ungrounded Specimen Test (UST).

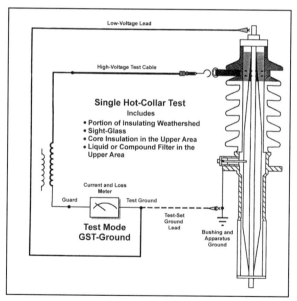

Figure 4.10 – Illustrating the Hot-Collar Test method for testing of bushing insulators (courtesy of Doble Engineering Co.).

guideline, losses up to 0.10 watts at 10 kV and 6 milliwatts at 2.5 kV are considered acceptable. External contamination and moisture will affect the measured losses. Therefore we recommend that bushings be clean and dry before testing.

Liquid Insulation Power Factor Test

To complete the routine power factor testing of the transformer insulation system, draw a sample of oil from the transformer and test for power factor. Although this test alone reveals little about the remaining life expectancy of the oil, properly applying the results to other oil tests will assure you that the oil is fit for continued service.

New oil, according to ASTM D 3487-95, should have a power factor of 0.05% or less at 20 °C. Power factors for in-service oils can gradually increase to as high as 0.5% at 20 °C. However, the power factor exceeding 0.5% calls for an investigation. Deciding how to treat the oil depends on what is causing the high power factor. Karl Fischer moisture content tests can indicate if moisture is the cause of the high power factor.

Single-Phase Excitation Current Tests

Single-phase excitation current measurements for field detection of shorted turns and core damage was developed in 1967. This test method is a natural extension of the power factor test and makes use of the same test equipment. Numerous cases of extensive core problems, like shorted lamination or winding problems, partial or high-resistance short-circuit between winding turns, have been detected using single-phase excitation current measurements.

The following guidelines should be followed for routine excitation current tests:

- Disconnect all loads and de-energize the transformer.

- Exercise caution in the vicinity of all transformer terminals because voltage will be induced in all windings during a test.

- Routine tests can be confined to the high voltage windings. Defects in the low voltage windings will still be detected and the excitation current required will be reduced.

- Winding terminals normally grounded in-service should be grounded during tests except for the particular winding energized for the test. For example, with a wye/wye transformer, the neutral of the high voltage winding would be connected to the UST (Ungrounded Specimen Test) circuit, while the neutral of the low voltage winding would be connected to ground.

- For routine tests the load tap changer (LTC) should be set to neutral, then to one step above neutral, then to one step below neutral, then to full raise or full lower. To ensure that the tap selector is functioning properly throughout the entire range of selection, you may want to perform tests on all LTC positions.

- Test voltages should not exceed the rated line to line voltage for delta-connected windings or rated line-to-neutral voltage for wye-connected windings. Generally, these tests are made at 2.5, 5, or 10 kV as the capacity of the test equipment permits.

- Test voltages should be the same for each phase. Because of the nonlinear behavior of exciting current, test voltages should be set accurately if results are to be compared.

- For single-phase transformers, excitation current test results are recorded with high-voltage windings energized alternately from opposite ends. This should also be done on the individual phases of three-phase transformers if the unit is suspect or if the initial exciting current measurements are questionable.

Figure 4.12 illustrates the excitation current test procedure for single-phase transformers. Figures 4.13 and 4.14 illustrate the excitation current test procedures for three-phase wye and delta connected units respectively. Note that all excitation current tests are performed by the Ungrounded Specimen Test (UST) method.

On single-phase transformers, the two currents obtained should be the same. Currents recorded for single-phase transformers should be compared either with similar units or with data obtained from previous tests on the same unit.

The test results recorded on the individual phases of three-phase transformers are also compared. For a three-phase core form transformer, a pattern of two similar currents and one low current is expected. The lower current for one of the phases, usually phase H_2-H_0 for wye-connected windings and phase H_1-H_2 for delta-connected windings, is associated with that winding which is wound

Figure 4.11 - Liquid-insulation cell connected for ungrounded specimen testing (courtesy of Doble Engineering Co.)

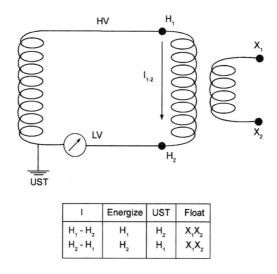

Figure 4.12 – Measurement of I in a single-phase transformer.

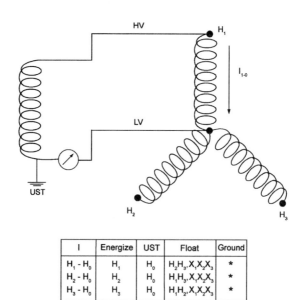

Figure 4.13 – Measurement of I in a wye-connected transformer winding (routine method).

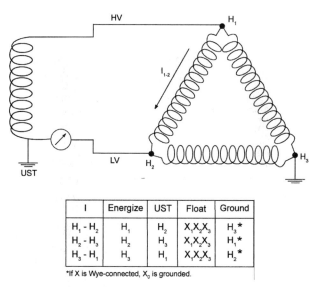

I	Energize	UST	Float	Ground
$H_1 - H_2$	H_1	H_2	$X_1X_2X_3$	H_3*
$H_2 - H_3$	H_2	H_3	$X_1X_2X_3$	H_1*
$H_3 - H_1$	H_3	H_1	$X_1X_2X_3$	H_2*

*If X is Wye-connected, X_0 is grounded.

Figure 4.14 – Measurement of I in a delta-connected transformer winding.

on the center leg of a three-legged core. The magnetic reluctance of this phase is lower than the other two phases and results in a lower excitation current value. Three-phase shell-form transformers may also yield a pattern of two similar currents and one low current, but only if the secondary winding is delta-connected.

If single-phase excitation current tests were included in the factory test specifications, comparing test data reveals changes undergone between the factory and the field.

On repaired transformers, the value of excitation current will generally increase along with, but not necessarily proportional to, an increase in core loss. Joint construction severely affects the magnitude of the excitation current. Changes in the hysteresis and eddy current characteristics due to handling the steel also affect the excitation current.

High precision in excitation current measurements moreover, does not appear to be necessary. The serious faults found have increased excitation current magnitudes by greater than 10% over normal values.

Megohm Meter (Insulation Resistance) Series

The insulation resistance measurement on equipment in service is one of the oldest techniques in testing. This resistance test and others in this series of electrical tests are usually measured by a specially constructed meter with both current and voltage coils and a meter calibrated to read directly in ohms or megohms. This megohm or Megger® meter has a built-in direct current generator that may be either hand-cranked or electrically driven to develop a high dc voltage that causes a small current to flow through and over the surfaces of the insulation being tested.

The total current through and along the insulation is made up of three components:

1. Capacitance Charging Current

   Current that starts at a high value and decreases after the insulation (capacitance) has been charged to full voltage.

2. Dielectric Absorption Current

   The current component resulting from absorption within imperfect dielectrics is caused by various polarizations. The most predominant polarization is the interfacial type resulting from the barrier effect at the interfaces of materials within such composite structures. A dipole polarization resulting from polar molecules and molecular chains will also be found at the DC end of the spectrum in some types of insulation. This energy causes voltage to reappear at the electrodes or plates of a dielectric after the stored energy resulting from the capacitance has been dissipated by a short circuit and the short circuit removed. Because of this fact, the short circuit should be held on the insulation under test for at least as long as the total test time or longer.

3. Leakage Current

The leakage current is the most important component in evaluating the condition of insulation. The path of this current may be either through the volume of insulation or over leakage surfaces. This component, unlike the other two, represents a loss current. Theoretically, the leakage current should be constant over time for any one value of applied voltage. This constancy over time indicates that the insulation under test can withstand the applied voltage. Any tendency for this current to steadily increase with time at a constant applied voltage is a warning that the insulation may be damaged if the test is continued at that voltage.

The insulation resistance measurement can be of value in determining the presence or absence of harmful contamination or degradation. However, its marked sensitivity to relatively small changes in the apparatus being tested can result in confusing data. As a result, pay attention to all possible variables in order to reach a sound and safe interpretation of the test results. Such variables include the effects of temperature, humidity, and external leakage due to dirty insulators and bushings, duration of the test, etc. The temperature of the equipment under test, for example, has a marked influence on the results obtained. Insulation resistance values decrease with increasing temperature. Therefore, if you want to reliably compare readings from different time periods, these readings should be taken at approximately the same temperature. This may be difficult to do. If readings are taken at different temperatures, all readings should be corrected to a common base of 20 °C. A short listing of temperature correction factors is shown in Table 4.1.

Figure 4.15 – Megger® Insulation Resistance Test Set (courtesy of Megger, formerly AVO International).

This test is relatively quick and easy. Insulation-resistance measurements are taken from the individual windings to ground and winding to winding. The tests consist of applying voltage to a winding for a period of 10 minutes.

We advise you to watch the trend of successive readings as well as the minimum values. A downward trend over a period of time may reveal changes that justify investigation.

Temperature Correction Factors for Oil-Filled Transformers

Base Temperature 20°C		
°C	°F	Correction Factor
0	32	0.25
5	41	0.36
10	50	0.500
15	59	0.720
20	68	1.00
30	86	1.98
40	104	3.95
50	122	7.85

Table 4.1

Minimum Insulation Resistance

You can also determine what the condition of the insulating system is if the measured resistance is above a calculated minimum resistance value. This is done mainly for transformers and is calculated prior to the series of tests. This figure then becomes a reference point by which the one-minute measured resistances may be compared.

From the information on p. 205, calculate the minimum insulation resistance values. Insulation resistance values near or below these calculated minimum values point to probable trouble. Corrective maintenance may be required.

Figure 4.16 represents typical insulation resistance test data. Readings have been recorded at 15, 30, 45 and 60 seconds and then every minute thereafter for a total 10 minutes. In Figure 4.16 the test voltage of 2500 V was connected Primary/Ground, Primary/Secondary, and Secondary/Ground. Now consider an evaluation of the test data.

$$R = \sqrt{\frac{CE}{kVA}}$$

Variables:
    kVA = Rated capacity of winding under test
    R = Insulation resistance of winding to ground with other winding grounded or between windings in megohms at 20 °C (or)
    R = kV + 1, where kV equals voltage classification of the winding under test (for transformers rated less than 100 kVA)

Testing without guard terminal:
    C = 0.8 for oil-filled transformers at 20 °C (or)
    C = 16.0 for dry, compound filled or untanked oil-filled transformers
    E = Voltage rating of winding under test

Testing with guard terminal:
    C = 1.5 for oil-filled transformers at 20 °C (or)
    C = 30.0 for dry, compound filled or untanked oil-filled transformers

These apply to single phase transformers or to separately tested windings of multphase units.

If the transformer is the three-phase type and the three windings are being tested together, then:

    E = Voltage rating of one of the single phase windings (phase to phase for delta connected units and phase to neutral for wye connected units)

Dielectric Absorption

Readings can be plotted on log-log paper. The resistance is coordinated with time and the resulting curve for a good insulating system will display a rise for the duration of the test (Figure 4.17).

Polarization Index (PI)

From the dielectric absorption data, the polarization index can be determined (Figure 4.17). Polarization index, the ratio of the 10-minute resistance to the 1-minute resistance value, is a dimensionless quantity often used in insulation evaluation. It is a numeric ratio and offsets the fact that previous test information may not be available. Since the leakage current increases at a faster rate with contamination or damaged insulation than absorption current, the polarization index will be lower for insulation in poor condition. Since all the variables that can affect a single Megger® reading are

Power Transformer							
Company - S.D. Myers, Inc. Location - Tallmadge, OH						Test No. 1	
Equipment - **3 Phase Transformer** Type - **FOA**			Rating - **50 MVA** Mfr. - **GE**			Voltage - **69/13.8 kV** Winding Age - **1956**	
State if Transformer is Oil-Filled and of the Conservator, Free-Breathing or Gas-Filled Type					**Conservator**		
If of the Free-Breathing Type, describe the Breather Arrangement and Dehydrating Medium Used							
	Test Data - Megohms				CF = 1.98 Top Oil Temp 30°C		
Test Voltage	2500 VDC		2500 VDC			2500 VDC	
Test Connections	PRI GND	To Line To Earth	PRI SEC		To Line To Earth	SEC GND	To Line To Earth
1/4 min	450		525			400	
1/2 min	475		625			425	
3/4 min	500		700			450	
1 min	550		725			480	
2 min	650		850			600	
3 min	700		925			625	
4 min	725		975			650	
5 min	750		1000			660	
6 min	775		1050			670	
7 min	775		1100			680	
8 min	790		1125			690	
9 min	790		1150			700	
10 min	800		1150			700	
10/1 min ratio	1.46		1.59			1.45	
Remarks: Step Voltage Test @ 500 VDC for 1 min = 400M ohms/SEC Step Voltage Test @ 500 VDC for 1 min = 500M ohms/PRI							

Figure 4.16 - Typical insulation resistance/dielectrc absorption test data.

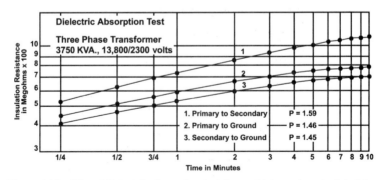

Figure 4.17 – Plot of dielectric absorption test from which can be calculated the PI (polarization index) (courtesy of Factory Mutual).

essentially the same for both readings, information of real value can be obtained. The PI Table 4.2 lists the parameters to which an insulating system can be categorized.

Step Voltage Test

The step voltage test determines the presence of moisture in the equipment's insulation. It is rated at voltages equivalent to or greater than the highest voltage available from the insulation resistance tester. Even though the highest voltage available does not stress the insulation beyond its rating, a two-voltage test can often reveal the presence of harmful contaminants.

The applied one-minute voltages should be in a ratio of 1 to 5 or greater: for example, 500:2500 or 1000:5000. Results showing a decrease of the insulation resistance at the higher applied voltage usually will indicate the presence of contamination in the insulation system. The rule of thumb is that a 25% or greater difference is cause for concern. Since weak insulation will display a lower resistance at the higher applied voltage, note any difference during evaluation.

Polarization Index Guide for Evaluation

Condition	Polarization Index (PI)
Dangerous	Less than 1.0
Poor	1.0 - 1.1
Questionable	1.1 - 1.25
Fair	1.25 - 2.0

Observe the Polarization Index calculation in Figure 4.17

Table 4.2

Transformer Turns Ratio Test (TTR)

The turns ratio test is primarily used as an acceptance test. We also use it as a tool for investigating problems, as well as an integral part of a routine preventive maintenance program.

During the manufacture of new transformers, the turns ratio test is performed on all tap positions to verify that the internal connections are correct and that there are not short circuited turns. It is listed as a design test according to Std C57.12.00-2000."Standard General Requirements for Liquid Immersed Distribution, Power, and Regulating Transformers"

During routine maintenance tests, the turns ratio test should be performed on all tap positions to identify short-circuited turns, incorrect tap settings, errors in turn count, mislabeled terminals and failure in tap changers. If the transformer has been modified or repaired in the field, or if a fault has occurred dropping the unit off the line, a TTR test successfully checks the integrity of the transformer windings.

The voltage rating or ratio you see on the transformer nameplate is determined by the number of turns of wire on the primary and secondary windings. Therefore, the turns ratio between the primary and secondary is equal to the voltage ratio of the primary and the secondary. The turns ratio does not tell how many turns of wire are on the primary or secondary coil, but only gives the ratio.

The ratio test can be made either of two ways. One method involves applying a known voltage on one winding and measuring the induced voltage on the other winding. On three-phase transformers connected in wye with an external neutral, the ratio can be checked with a single-phase potential. The manufacturers recommend that a voltage of at least 10% of the rated voltage be used when making the ratio test by this method. However, this is not always a practicable method to use in the field.

The most well known method of checking turns ratios in the field utilizes the transformer turns ratio test set. (Figure 4.18). This test set has an internal generator to supply the test potential to a reference transformer in the test set and the low voltage winding of the transformer under test. The measured ratio should compare with the calculated ratio on any given tap to within ±0.5%.

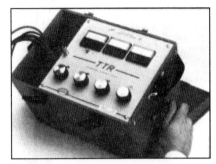

Figure 4.18 - A transformer turns ratio (TTR) test set (courtesy of Megger, formerly AVO International Co.)

A turns ratio measurement can show that a fault exists but will not determine the exact location of the fault. An internal inspection or detanking may be needed to locate the problem.

A typical set of turns ratio test date would be as follows:

Transformer connected delta-delta
Primary voltage: Tap #3= 12470 V
Secondary voltage = 4160 V
Calculated turns ratio = $\frac{12470}{4160}$ = 2.997

Measured ratios:
H_1-H_2 / X_1-X_2 = 2.996
H_1-H_3 / X_1-X_3 = 2.995
H_2-H_3 / X_2-X_3 = 2.999

These measured ratios are all within the ± 0.5% of the calculated ratios

Transformer connected delta-wye
Primary voltage: Tap #3 = 12470 V
Secondary voltage = 4160/2400 V

Calculated turns ratio = $\dfrac{12470}{2400}$ = 5.196

or

$\dfrac{12470}{4160}$ = 2.997 × $\sqrt{3}$ = 5.191

Measured ratios:
$H_1\text{-}H_2 / X_0\text{-}X_3 = 5.194$
$H_2\text{-}H_3 / X_0\text{-}X_3 = 5.197$
$H_3\text{-}H_1 / X_0\text{-}X_1 = 5.195$

These measured ratios are all within the ± 0.5% of the calculated ratios:

Transformer connected wye-wye
Primary voltage: Tap #3 = 12470/7200 V
Secondary voltage = 4160/2400 V
Calculated turns ratio = $\dfrac{7200}{2400}$ = 3.000

Measured ratios:
$H_0\text{-}H_1 / X_0\text{-}X_1 = 3.009$
$H_0\text{-}H_2 / X_0\text{-}X_2 = 2.990$
$H_0\text{-}H_3 / X_0\text{-}X_3 = 3.011$

These measured ratios are all within the ± 0.5% of the calculated ratios.

DC Winding Resistance Test

The DC winding resistance made with a low resistance ohmmeter or sometimes referred to as a DLRO® (Digital Low Resistance Ohmmeter) will indicate a change in DC winding resistance when there are short-circuited turns, poor joints, bad contacts, and changes in the windings due to a change in capacitance. When possible, the readings should be compared to the factory test data or previous field test data. If previous factory or field data is not available, use the data to compare phases on a three-phase unit or another identical unit.

The low resistance is measured by the potential drop method. The low resistance ohmmeter has a high range scale of 0 to 5 ohms; the lowest range measures from 0 to 500 microhms. It has accuracy within one percent of full-scale deflection on all ranges. As current is applied to the winding, the resistance will be constant, so the potential will show any voltage drop present. Readings should be in the milliohm range. Prior readings are necessary so that a comparison can be made. High readings can indicate a problem with bonded, bolted or spring-loaded connections that if left unattended may cause problems. If winding deformation takes place, it will affect the capacitance, and in turn, affect the winding resistance.

On three-phase transformers, the measurements are made on the individual windings (phase to neutral) when possible. On delta connections, there will always be two windings (in series) which are in parallel with the winding under test. For this reason, on a delta winding three measurements should be made to obtain the most accurate results.

Since the resistance of copper and aluminum varies with temperature, all test readings must be converted to a common temperature to give meaningful results.

Most factory test data are converted to 75 °C which has become the most commonly used temperature. The formula for converting resistance readings on copper winding (to 75 °C) is as follows: (C = 234.5 for copper, C = 225 for aluminum)

$$R\ 75\ °C = (R\ test) \times \frac{234.5 + 75}{234.5 + WDG\ temp\ in\ °C}$$

Because the winding resistance test cannot measure temperature precisely, the permissible deviation for this test in the field is two percent of the factory test values.

The DC winding resistance test is also applicable to circuit breaker contacts, tap changers, coils, and other electrical devices.

DC Overpotential Testing

Some authorities claim that DC overpotential testing is a potentially destructive insulation test because it stresses solid insulation more than the oil. You must weigh the risk before proceeding with the test. You can find information on the recommended maintenance test voltages in the following:

- Std. C57.12.90 1993, *IEEE Standard Test Code for Liquid-Immersed Distribution, Power and Regulating Transformers.*

- International Electrical Testing Association Inc. (NETA) Maintenance Testing Specifications NETA MTS-2001

- Insulated Cable Engineers Association (ICEA)

- National Electrical Manufacturer's Association (NEMA).

If published information for determining the direct-current voltage values needed during maintenance testing is not available, an alternate method of calculating the DC test voltage may be used. A figure commonly used in maintenance testing of in-service equipment is twice the rated voltage plus 1000 for AC tests. For DC testing, as discussed here, the figure is then multiplied by the AC to DC conversion factor of 1.6. This voltage is then derated by 65% for older equipment. Hence, the formula for maintenance testing of older equipment is:

$$ET = (2\ EN + 1000) \times 1.6 \times 0.65$$

ET = maximum DC test voltage
EN = nameplate voltage
1.6 = AC to DC conversion factor
0.65 = derating factor for older equipment

If you compare leakage current from phase-to-phase or with previous, you'll get an idea of the insulation's condition. Generally, higher leakage current indicates poor insulation. However, the latter is difficult to evaluate from a single test. As long as the curve of "Leakage Current vs. Voltage" is of the general shape shown in Figure 4.19 and as long as the knee occurs at a sufficiently high voltage (above the maximum test voltage required), you can be satisfied with the tested equipment.

Core Ground

Another test in the field is the core ground check, which is perormed with an insulation resistance test set. This test is most important as a check for shipping damage. So consider doing the core ground check routinely.

Here's how it works. Heating will occur in a core form transformer due to circulating currents if the core laminations are grounded at more than one point. The laminations that make up the core are insulated from each other and from the end frames and other structural members. These structural members, in turn, are normally connected to the main tank. The laminations are insulated from each other by a chemical treatment that leaves a very thin layer of insulation. This insulation between the laminations

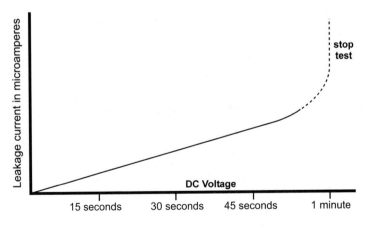

Figure 4.19 - Evaluating DC overpotential test results.

is only a few ohms of resistance, but sufficient to prevent damaging eddy currents within the core.

The core ground connection is usually located at the top of the transformer. On some designs, this connection is not solid but instead is made through a heavy-duty resistor in the 250 to 1000 ohm range. A resistor in this range still accomplishes the effective grounding of the core and at the same time limits circulating currents. This ground connection is usually conveniently mounted under a manhole at the top of the transformer. Some designs bring the connection through the tank with a small low voltage bushing and the grounding strap is external. These accessible locations permit the ground to be easily removed and thus the core-to-ground insulation may be readily tested.

The core-to-ground insulation normally consists of a pressboard material that should withstand at least 2000 volts. Damage incurred in shipment can very likely affect this core-to-ground insulation. Therefore, it is recommended that it be tested before unloading the transformer from the railroad car or truck on which it was shipped. Using a 1000-volt Megger®, this insulation generally should read 200 megohms or more between ground strap and end frame.

If the Megger® indicates a ground between ground strap and end frame, switch to an ohmmeter and read the resistance. If the resistance reading is only 1 or 2 ohms, it can be considered a solid ground and the unit should be repaired. If the resistance is 200 to 400 ohms, it is considered an inadvertent or high resistance ground. There are various field methods available to remove these inadvertent high resistance grounds. One method is explained in Standard 62-1995 *IEEE Guide for Diagnostic Field-Testing of Electric Power Apparatus - Part 1: Oil Filled Power Transformers, Regulators, and Reactors.*

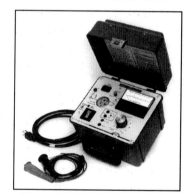

Figure 4.20 – High Pot Tester (courtesy of Megger, formerly AVO International Co.)

Ground Resistance Measurements

Ground resistance is the resistance in ohms between a grounding electrode and a remote or reference electrode.

You may wonder why ground resistance is of much concern in a substation or any power system network. After all, one can see that the transformers all have copper conductors attached to their respective metal bases and the conductors extend down into the earth. It would appear that all the units are properly grounded, right? Not necessarily so! What has happened to the copper conductors beneath the soil? The grounding electrode or ground grid may have corroded sufficiently to increase the ground resistance to an unsafe level. If the equipment (transformer, tower, etc.) is not properly grounded, under certain conditions dangerous potentials could exist between the equipment and ground. A person touching the equipment would then be subject to this potential. Therefore, ground resistance in substations and power system networks is a vital concern to plant maintenance personnel and should be measured periodically to see that it remains at a safe, low value.

One method to measure the ground resistance is the "fall-of-potential" method. The basic diagram is shown in Figure 4.21.

Several manufacturers supply equipment that utilizes this method. Present day instruments provide a readout directly in ohms, thus eliminating any calculation. Although the newer equipment is more sophisticated, it still performs the basic functions for the fall-of-potential testing.

The soil condition, temperature, and moisture content of the soil have a great bearing on the value of resistance that is measured. A single ground rod does not always provide a sufficiently low resistance for large power systems. A ground grid consisting of a system of grounding electrodes interconnected by bare cables buried in the earth may have to be constructed to obtain the safe, low ground resistance. If the ground electrode or grid is known to be in good condition, but the measured resistance is high, the soil may need to be conditioned with water or chemicals.

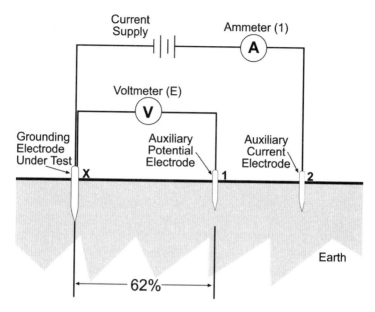

Figure 4.21 – Fall-of-potential method of ground resistance testing.

A good grounding system keeps the ground resistance to an acceptable low value. This condition serves to protect personnel and equipment. The National Electrical Code contains detailed rules for grounding. These rules specify that the grounding electrodes have a resistance to ground of not more than 25 ohms. However, it is recommended that the resistance be kept at 5 ohms maximum per IEEE Std. 142.

Leakage Reactance Test

Leakage reactance measurements are used to detect winding movement or displacement in power transformers. Often, a unit will remain in service with partially deformed windings; however, the reliability of that unit would be significantly reduced. As a useful test, troubleshoot transformers after an event like a short circuit. The leakage reactance measurements obtained compare to measurements obtained at the factory or recent field tests. If no recent tests are available, it may be difficult to ascertain if the damage was from a recent event or had accumulated over the years. Performing routine leakage reactance measurements is also important in monitoring coil movement. These tests may prove to be beneficial in detecting problems before a failure occurs.

Leakage reactance measurements can be made with the Doble M4000 insulation power factor test set using the M4110 Leakage Reactance Interface (Figure 4.22). The test consists of short-circuiting one winding of the transformer and applying test voltage to another winding. The magnetic flux path includes the space between and within the windings. Reactance measured under these conditions is determined by the reluctance of that space and is referred to as leakage reactance. Leakage reactance is sensitive to winding distortion and insensitive to temperature and contamination like water or particulate matter. The information obtained during the test can be compared to the short-circuit impedance nameplate value obtained at the factory.

Frequency Response Analysis (FRA)

Presently, several European utilities and engineering companies use frequency response analysis to obtain fingerprint values of new and used equipment. The information will then be used to investigate the mechanical integrity of the transformer at a later date. Ontario Hydro in Canada originally developed the method in the late 70's. It was then further advanced and refined in Britain and Ireland in the 80's and 90's.

The frequency response analysis method detects mechanical problems inside the transformer caused by events like short circuits or switching serge. This test in recent years has proved to be easy to perform and provides reliable information on the mechanical integrity of the unit. The results are helpful even if previous data is not available.

Forces on the transformer during short circuit can be hundreds of thousands of pounds! Transformers are designed to sustain these

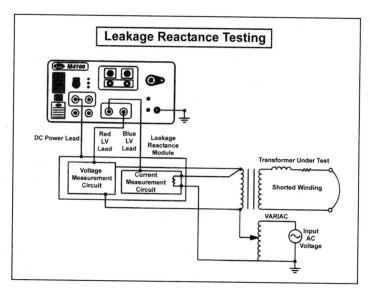

Figure 4.22 - M4100 Instrument and M4110 Leakage Reactance Interface Test Connections (courtesy of Doble Engineering Co.)

forces provided the mechanical strength of the unit has been maintained. Transformers are likely to withstand many short circuits during their service life. But sooner or later winding movement may occur and the ability to survive another short circuit will be compromised. It would be beneficial to test the unit periodically with FRA to obtain information that may indicate a problem.

There are two methods of FRA: the impulse method and the swept frequency method. Similar readings can be obtained using either one. The swept frequency method seems to be better adapted for site use. The swept frequency method also has superior signal to noise performance at the high frequencies considered important in detecting winding changes.

The test equipment consists of a network spectrum analyzer, a notebook computer, an interface, three 50-ohm coaxial test leads, and some grounding leads. The interface allows communication between the computer and the analyzer.

Recovery Voltage Method (RVM)

The RVM test provides information on the transformer insulation. Because RVM is a new test, we need experience to correct its performance and to evaluate its results.

When you apply a DC voltage to the insulation, molecules are polarized and turn in the direction of the electric field. A short circuit is then applied for a short period of time and the molecules are partially depolarized. Upon opening the short circuit, a voltage due to the remaining charge will build up between the terminals of the insulation. This is the recovery voltage.

From this information you can determine what the moisture content of the insulation is and the maximum allowable operating temperature.

Chapter 5

Transformer Coatings

Signposts of Paint Deterioration

Continuous surface inspection should focus on the following trouble indicators:

The following coating failures can be easily located in areas such as the bases of transformers, particularly where units sit on concrete without air vents; sharp edges such as cooling fins; angle iron bracings welded onto fins for rigidity; the lower third of cooling fins, close to the manifold; underneath of top-lid, and bottoms of radiators due to corrosives migrating. Coating failures of this nature need immediate attention, if deterioration is to be checked.

Chalking
The result of weathering of epoxy paint at the coating surface is known as chalking. Cause: sunlight. Contrary to "house paint" claims, controlled chalking can NOT be an asset as a self-cleaner on a transformer. It may even attract additional air-borne particulate matter.

Checking and Cracking
Describes the breaks in the paint film which are formed as the paint becomes hard and brittle. Major cause: temperature variation (expansion and contraction of the steel surface).

Flaking and Peeling
As penetrating moisture loosens relatively large coating areas, the paint curls slightly, exposing more of the basic surface, and finally flakes off. Cause: wrong choice of paint or improper cleaning of surface.

Mildew
> A fungus grown in the presence of moisture which feeds on the coatings. Factor generally only associated with house paints.

Rust (Corrosion)
> Deterioration of a substance (usually a metal) because of reaction with its environment; especially with oxygen, moisture and the oxides of carbon, nitrogen, and sulfur (ASTM D 610-95 Rust Guide).

Permeability
> The process of soaking up water by the paint film and the transmission of the water through the metal surface. Then, hydrogen (which is at the surface) develops embrittlement.

Osmosis
> The accelerated blistering of the paint film due to the formation of soluble salt solutions beneath the coating caused by moisture (water vapor transmission - W.V.T.).

Spalling
> Separation or coat loosening adhesion and peeling away. Paints won't tolerate excessive surface temperature.

Solution to Paint Deterioration

The application of protective coatings to steel surfaces affords a solution to the corrosion problem. Such solution involves a thorough preparation of steel surfaces, followed by the application of primer and finish coats.

Preventive Versus Remedial Painting

Traditionally, transformers have been the most neglected piece of equipment in the plant. Today, they are among the easiest to maintain and to protect, and the hardest to replace. Keep in mind that the condition of the coating and surface is undoubtedly the most important in any repair or maintenance program.

Preventive painting begins at the first phase of deterioration, that is before rust begins to appear. The longer the decision to repaint is put off, the higher the costs. If the decision is put off too long, you are playing roulette with your electrical equipment - and finally oil leaks in the cooling radiators. In fact, you are into remedial painting, which costs a minimum of 25% more than preventive painting.

Three basic questions should be answered concerning an on-going maintenance program:

1. What is the condition of the original coating, and how much disintegration or corrosion has already occurred?

2. What is the condition of the basic surface? Both the type of repair and the type of coating vary with the underlying surface condition.

3. What is the existing type of coating on the surface?

If the ASTM Standard (ASTM D 610-95), which was developed to rate the condition of a surface is recognized and followed, the condition of the surface can be rated by several people. The evaluations of paint deterioration should be reasonably consistent.

The first phase of such deterioration normally shows up as changes in the top coat appearance with no significant effect on its protective qualities. These changes include a combination of soiling, color change, loss of gloss, rust, bleeding, or chalking. The vital signposts of paint deterioration are briefly described in the next section.

Surface Preparation

Correct surface preparation is the key to the lifelong performance of any coating system. Studies have shown that about 80% of all coating failures are a result of inadequate or incomplete surface preparation.

Quality surface preparation can be costly, but one should never reject it because the longer repainting interval will offset the additional cost. Since labor is the major cost factor, comprising approximately 70% of the job price, it is much cheaper to do it right than to do it all over.

There are three principal methods of preparing steel surfaces for painting.

Scraping and Wire Brushing

Scraping and chipping of very thick layers of old paint and other contaminants are often used as a preliminary to other preparation techniques such as wire brushing or power tool cleaning. Wire brushing or hand tool cleaning is one of the most widely used techniques for removing loosely adhering mill scale, deteriorated coatings, and rust. Unfortunately, wire brushing does not usually reach lower contours of the steel, except the upper contours of surfaces. Thus, sound coatings which are glossy or very hard to clean should be "sweet-blasted" (very light sand blasting) to slightly roughen or "pickle" the surfaces. This will help maximize the adhesion capacity of the new coating. Wire brushing equipment is varied, and includes a wide selection of brush sizes and shapes made from different types of bristles.

Removal of heavy rust scale or previous coatings over extensive areas can be done by electrically powered equipment. Power tool cleaning usually results in a better job compared to hand tool cleaning. This technique is much faster, easier, and more economical.

Powder Blasting

Note: Before considering powder blasting the paint on any equipment or structures, ascertain the impact of all federal, state and local regulations regarding containment and disposal of lead-based paints.

Blasting is perhaps the most effective method of removing mill scale, old paint, rust, hardened oil, and other surface layers of the steel substrate. This can be achieved by blasting with a mixture of pulverized limestone and/or corn cob. Limestone has a Rockwell hardness rating of 4.0, so it will not damage the metal or porcelain. Sand is not recommended since it will damage porcelain and is too harsh for cooling radiators or glass insulators.

For best results, the following sequence of operations should be followed:

1. Determine if you are dealing with an item that has been previously painted with a lead-based paint. If so, proper containment of all blasting debris as well as OSHA safety requirements in handling lead-based paints must be followed. This may be quite costly and alternatives such as encapsulating or disposal may be another choice.

2. Before blasting, remove spilled oil, salt, chemicals, dust and other contaminants by solvent wash.

3. Before blasting, remove hardened oil and grease by sprayings with a chemical degreaser to soften. Caution: Rust forms under the oil and leaks may have already developed prior to surface preparation.

4. Blast to the desired grade such as: white metal blasting in which all mill scale, rust, rust scale, previous coating, etc., are completely removed, leaving the surface a uniform gray white color; and near-white blasting in which the metal is cleaned until at least 95% of each square inch of surface area is free of all visible residues.

5. After blasting, all grit and dust should be removed with a clean brush, vacuum cleaner or dry and clean compressed air. Before using a primer coat, all areas close to the surfaces to be painted should be wetted down to prevent wind-blown debris from contaminating the paint.

6. In areas of severe corrosion, solvent wash the substrate with mineral spirits or naptha reducers.

7. Painting of the blasted surfaces should be done as soon as possible, that is, before contaminants settle on the cleaned surface.

Even with proper surface preparation, it is safe to predict that rust will recur again in the same areas. What then can be done? Years of experience have led one major contractor to apply a special pre-primer to help neutralize rust in such uncontrolable areas. Pre-primers are helpers only.

More severe rust-pitting may result in oil leaks if mechanical blasting is used (Figure 5.1). Cooling radiators are particularly thin and subject to greater corrosion from within (by oil decay products) and from without (by atmospheric pollution and weather) (Figure 5.2). Also, radiators do not have as much internal heat to dry the surface as does the main tank. As a matter of interest to show how thin some of the radiators are, many reported cases show that they will actually collapse under only four psi vacuum.

Other important considerations in abrasive blasting include the type of equipment used, such as nozzles and pressure hoses.

Figure 5.1 - Rust-pitting

Figure 5.2 - Rusty pancake radiators.

The size of nozzles and distance from substrate are also important as they can affect the quality of the finished blast. During the blasting process, operators should wear a protective facemask to protect themselves from the hazards of fine dust. To avoid such hazards, wet blasting is recommended. This involves the use of water-soluble inhibitors, such as phosphates and chromates - which can effectively retard superficial corrosion after sandblasting.

Chemical Cleaning (De-Energized)

Chemical cleaning or acid pickling is probably the oldest technique of preparing surfaces for a protective coating. The method involves the use of dilute acids (up to ten percent) such as sulfuric, phosphoric and hydrochloric solutions. This method is seldom used as a field paint stripping system. Containment is difficult and with lead-based paints, ground contamination is a concern.

The Primer Coat

Primers are important insofar as exterior finishes are concerned because of the exposure of the top coat to the outside elements, to variations in temperature and humidity, and to internally generated heat.

Primers play a major role in protective coatings. First, a selective primer acts as a glue, providing an adhesive bond between the prepared surface and the finish coat. Universal primers will not lift existing aged paint, but will accept high performance coatings such as epoxies and polyurethanes. Lack of primer adhesion will result in the coating system peeling off in "sheets." Second, a primer acts as a rust inhibitor. A passive rust-inhibiting primer helps to neutralize hidden rust, control under film cutting, and adds strength to the coating system. Typical rust inhibitors include zinc chromate, iron, aluminum dust, barium metaborate, and zinc dust.

Oil-based primers such as tung oil- and linseed oil-based primers contain one or a combination of metallic rust inhibitors. These metallic inhibitors like zinc and flaking aluminum also form heat barriers. Since transformer oil may ruin any product containing zinc, zinc-rich paint is not advised for painting pad-mounted electrical apparatus. Note that oil-bases primers are slow-drying, and therefore require a longer waiting period before the finish coat is applied. Of course, the warmer the apparatus, the faster the drying period.

Third, the primer must be compatible with the prepared surface, as well as the top coat, to assure top-coat adhesion. As examples, in non-corrosive or mild environments, oil-lead, phenolic, or epoxy ester primer may be used in conjunction with a silicone alkyd finish coating. In corrosive areas, vinyl alkyd (heavily rust-inhibitive pigmented), or epoxy primer may be employed along with an active solvent finish coat (that is, epoxies). Finally, in aggressively corrosive areas, a high build epoxy primer coupled with a Class V urethane finish may be chosen.

On the other hand, use of vinyl, zinc-pigmented, or chlorinated rubber primers should be avoided. A vinyl primer requires a bright metal surface, and has poor bonding (adhesion) qualities. Zinc-pigmented primers, as mentioned earlier, may be attacked by transformer oils. In addition, they are high build, very heavy (about 26 pounds per gallon), cannot be flow-coated and may add a lot of weight on the cooling radiators resulting in the impairment of heat transfer (internal to external). Chlorinated rubber primers are limited by temperature exposure (no more than 55 °C), as well as poor adhesion and cohesion (the attraction between atoms and molecules within a substance). Even the best of today's paints - urethanes - are difficult to apply in damp, humid areas.

Other conditions that dictate the final composite choice include:

1. Recoat time - the time after an application when a coating achieves maximum hardness.

2. Basic surface condition

3. Chemical and abrasion resistance

4. Type of application, such as:

    a. Immersion in liquids - continuous direct exposure to various liquids.

    b. Marine and chemical environment - underfilm corrosion because of affinity of coating to moisture, oxygen, or chemicals.

    c. Supplementary pigments - such as for the reduction of water vapor transmission that leads to blistering.

Thus, it can be realized that painting transformers is a unique service, requiring coating specialists who are familiar with the types of primer and finish coats. The next section deals with the third and final step in a coating system.

The Finish Coat

The finish coat performs several functions.

1. Aesthetics

   This is a very important area, because the first reaction to a paint job is "How does it look?" The use of high gloss coatings is functional as well as decorative (Figure 5.3).

2. Safety

   Figure 5.3 - High-gloss finish coat.

   Color coding, indicating various standard locations and hazards, as well as increased visibility. For example, some painting manufacturers use yellow to designate caution against physical hazards, and purple to designate radiation hazards (OSHA color coding).

3. Protection

   a. Weather: sunlight (U.V.), wind, abrasion pollution, heat, and humidity. For example, in winter an energized unit could be hot at the top, very cold at the bottom, with quite a differential between internally generated heat and external temperature (thermogalvanic corrosion). For comparative purposes, consider how much heat difference various coating systems will accept. High, continuous temperature tolerances are found in alkyd silicones (150 °C) and catalyzed epoxies (105 °C), polyurethane (90 °C), while low temperature tolerances are found in chlorinated rubber (60 °C) and vinyl (60–66 °C).

b. Water: dampness may stain, harm or destroy a coating. Keep in mind that water from industrial cooling towers has a low pH factor. Therefore, this water will capture corrosive droplets and contain them on the apparatus surface long enough to promote electrochemical reactions.

c. Chemical Environment: acids and alkalines like sulfuric, hydrochloric, nitric, phosphoric, formic, caustic soda, lye, lime, as well as fertilizers, tree sprays and road salt will attack any paint system. Therefore, by choosing the right paint system, longer life for the transformer protective coating can be assured, even under the most adverse conditions. Nevertheless, under severe conditions no paint is assured of longevity. ANSI/IEEE Standard C57.12.00-1993 states that, "Metallic flake paints such as aluminum, zinc, etc., have properties which increase the temperature rise of transformers except in direct sunlight."

In applying the finish coat, the following points must be taken into consideration. First, the finish coat must be compatible with the primer coat. Adhesion and cohesion are qualities of top priority. Several quality universal primers are available, and can minimize this problem. Second, tailor the coating system for the best chemical resistance by using catalyzed epoxy or ASTM Class V polyurethane. Third, if corrosive fallout or splashing exists, then any system will fail prematurely.

For instance, the most critical painting problems exist in chemical plants in and around the cooling towers. Two alternatives are possible. Either paint when the cooling tower is not operating, or build a covering for the transformer while various phases of painting are going on. Only a few minutes exposure of the prepared surface to over-spray before the primer is applied programs the paint system to failure.

As you think about color, think ASTM Class V urethane, for it exhibits the following qualities:

- Resistance to moisture permeation
- Temperature flexibility
- Corrosive resistance
- Ninety-five percent light reflection (white only) and excellent color retention (fade resistance).
- Hard finish. Fungus cannot start a growth in humid areas.
- Curing time (36 hours) when corrosive resistance takes effect. Note that the solids content may vary among different paint formulators. The foregoing criteria apply to ASTM Class V, catalyzed TDI (isocyanate) with ultraviolet absorbers (or HDI modified with acrylic resins).

Fourth, finish coating need not be thick to be effective. Too much millage is detrimental; it leads to early flaking, adds excessive weight to the cooling fins, and impairs radiation, convection, and ventilation. Six to ten mills of paint is the accepted average.

Finally, the methods of applying the finish coating are a significant factor to the longevity of the paint system.

Methods of Applying Paint

There are three methods used for applying a finish coating to steel surfaces - brush or roller (hot stick), flow method, and spraying. These methods may be used separately or in combination. This choice is not a simple decision. The economics of the three methods is weighed against the suitability of the system to each type of application.

1. **Brushing**

   Brushing is perhaps the oldest known method of applying paint to power transformers and other accessories. Although the method is no longer used to any great extent (it is a very slow process), it is still employed extensively for spot touching. Roller application is used in conjunction with brushing. Its advantage over brush application is that it is much faster and can cover large areas within minutes. Rolling may be an option if containment of the overspray is not possible, and overspray could affect buildings on vechicles.

2. **Flow Method**

   This application has been used successfully for painting or repainting electrical apparatus in the field. It has proven to be one of the best methods for painting radiators. The method can be used on ONAN/ONAF transformers, requires minimum equipment, and is rather inexpensive. The equipment, as shown in Figure 5.4, might include improvised sheet metal pans with flared edges to hold access paint from the tank; a gear pump with a pressure adjustment system; and a neoprene rubber hose with a flattened aluminum tubing for applying the paint.

   The coating process itself is fairly simple. Before painting, accessories such as bushings, nameplates, gauges, and breathers should be covered with masking tape to protect them against paint splashes. Fan blades are usually stainless steel or tempered. Thus painting them is not recommended. Where the technique is used by some utilities, painters have found it practical to use brushes for painting the cover and base of transformer units.

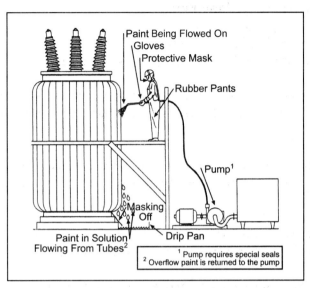

Figure 5.4 - Flow Coating Method

3. Spraying

High-pressure (or airless) spray application is considered a very effective process for coating electrical apparatus. Spraying equipment is both complex and costly, and ranges from simple aerosol spray bombs that can be purchased from hardware stores to high-pressure gun assemblies. The latter technique offers many advantages such as speed and efficiency, and allows the painter to get close to bushings, gauges, and valves. A drawback to this method is the inaccessability of some radiators, making it difficult to apply a uniform coat when spraying.

Where spraying is chosen for coating steel surfaces, external mix spray guns are often recommended. Before any spraying is done, all accessories should be masked to protect them against unnecessary coating.

Precautionary Measures

In any high quality protective coating, it is important that precautions be taken to assure professional results.

First, do not insist on blast cleaning to white metal, especially in fin areas. Keep mil thickness of paint under ten, or expect heat buildup.

Second, it is always advisable not to purchase the cheapest paint, and to avoid vinyl, zinc-pigmented, or chlorinated coatings.

Third, avoid purchase of paint without guarantee of mutual compatibility between primer and finish coatings.

Fourth, do not paint apparatus using dark colors or aluminum paint.

Finally, painting should be done only when ambient temperatures are suitable.

Where painting is being performed near energized equipment, plant safety personnel should always be consulted.

Chapter 6

Bushing Maintenance

Introduction

Since their inception, the problem of transformer bushing and stand-off insulator contamination due to the natural deposits (such as early morning dew, salt fog in sea coast areas, and more recently, smog) and subsequent industrial pollution has plagued the electrical industry. Such contamination has often resulted in noisy substations, damage to insulating surfaces, partial discharge, tracking flashover, and loss of power.

It is possible to prevent insulators from flashing over by periodic hand wiping (de-energized), periodic washing, or periodic dry cleaning using such techniques as dry powder blasting with an insulated "hot stick" or periodic coating with grease compounds (energized or de-energized).

The Contamination Problem

Bushing, insulator, and lightning (surge) arrester structures in the substation are subject to the same contamination problems as those that plague transformers and steel coatings; some authorities say they are even more so because of electrostatic attraction for airborne contaminations.

Build-up of contamination can, however, cause more problems to porcelain insulator structures than to metal parts of a transformer system. This is echoed by A.D. Lantz of Ohio Brass, who stated that:

> "Surface contamination and leakage currents caused operating difficulties eighty years ago... and exist today as one of the most serious insulator problems. Bushings and insulators are still not maintenance-free, and still need regularly scheduled cleaning to rid them of different types of contaminants."

Types of Contaminants

Contamination material comes in different sizes and varieties. Table 6.1 shows a range of typical pollution particle sizes.

Airborne contaminants can be divided into two general classifications: conductive and non-conductive or inert. Common conductive pollutants include carbon dust, fly ash, metallic particles, some ores, volcanic ash, and some chemical salts such as sodium chloride, sodium sulfate, magnesium chloride, and some acids. When in solution, these ionic salts can cause partial discharges and flashovers by providing a conductive coating on the surface of an insulator structure. The severity of pollution caused by these salts is generally measured in terms of the equivalent salt deposit density.

Non-conducting contaminants are composed of solid material that does not go into solution as ions. They include clays such as kaolin and bentonite, inorganic cements, silicone dioxide, as well as fertilizer and lime dust in farming areas. These inert contaminants may be hydrophobic, and thus increase the wetting rate of the insulator. Other inert materials such as grease and oil may also be hydrophilic, increasing the wetting rate of the insulator. Once grease and oils have reached their absorbent limits, the oils and grease harden and then support surface wetting. Non-conducting pollutants such as cement dust do not cause trouble when dry, but when the humidity is high, they absorb water and change from inert to conducting. When some critical value is reached, arcing over occurs.

Range of Typical Pollution Particle Sizes

Nature of suspended matter		Diameter (microns)
Inorganic	smokes	0.001 - 0.3
	fumes	0.01 - 1.0
	dust	1.0 - 100.0
Organic	bacteria	1.0 - 10.0
	plant spores	10.0 - 20.0
	pollens	15.0 - 50.0
Water	fog	1.0 - 50.0
	mist	100.0 - 100.0
	drizzle	50.0 - 400.0
	rain	400.0 - 4000.0

Table 6.1

Figure 6.1 shows the relative distribution of the most common contaminants. This data is based on seven years experience of a major transformer maintenance contractor and indicates that the highest percentage of service work has occurred in the areas of metallic and chemical contamination.*

Field experience has further shown that the worst contaminants are metallic oxides. Such oxides may even have a more destructive effect on transformer bushings because of the magnetic fields present (Figure 6.2). In extreme cases, internal heat coming up through the bushing from the transformer may increase the formation rate of glazed deposits. These two factors may help explain why the failure rate of bushings is much higher than that of standoff insulators.

Another form of bushing contamination comes from living creatures such as birds and rodents. Birds and rodent defecation can be an extreme nuisance for transformer owners and leads to very elaborate methods of prevention.

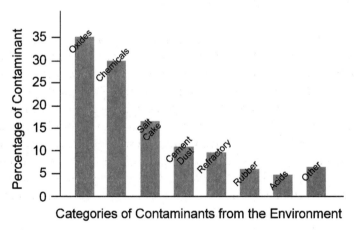

Figure 6.1 - Environmental Contaminants

*In addition, such sea-coast areas as the California and Gulf coasts, New Jersey and Maryland which frequently (daily in California) have salt fog but no rain for many months.

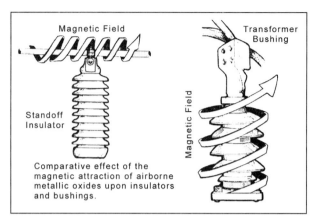

Figure 6.2 - Effect of metallic oxides on bushings/insulators. Bushings are even more critical because a bushing is a conductor, as well as an insulator. Thus, a magnetic field exists along the entire length of the bushing. Heat from the tank also compounds the rate of conducting potential.

Dynamics of Contamination Collection

1. Air movement

   Many researchers have investigated the dynamics of contamination collection on bushings and insulators. Results of such studies have not been conclusive on the factors that can be employed to accurately predict the collection rate.

   It is, however, generally accepted that air movement is responsible for the contamination of insulator structures. These contaminants can be easily observed in areas where wind direction is fairly constant.

   Most airborne particles are collected in areas of high stress such as the pin region on suspension insulators, where the electrostatic or electromagnetic field has some significant influence. This field also holds particles through a process of dielectric polarization once the particles are settled on the surface.

In industrial regions, especially near oil refineries and chemical plants, insulator surfaces may be subject to washing, such as heavy rain and an exposure to greater accumulation of wind-borne pollutants.

2. Moisture deposits

Moisture deposits on insulator surfaces are another way through which contamination is collected. Insulator surfaces can accumulate moisture through condensation, fog and impingement of water particles. Condensation takes place when the surface temperature of the insulator is reduced below the dew point temperature. Studies have revealed that flashovers on in-service insulators often occur in the early morning hours, the time when dew condensation is at its peak.

Impingement of water mist (droplets), which usually occurs in fog and rain, is another way in which contamination is collected on the under surfaces of insulators. Changes occur as droplets absorb chemical and metallic components. It should be noted that the direction of wind has a great influence on moisture deposition on insulator surfaces.

In cold areas, snow and ice have been observed to cause flashovers on insulators, particularly when they help to superimpose contamination. Studies have demonstrated that flashover voltage increases considerably before snow melts from the insulator surface, decreases as snow melts, and increases again upon complete disappearance of snow.

In coastal areas, the degree of contamination depends on distance from the sea. In other words, the closer to the seashore insulators are, the more serious the contamination, particularly for those areas that have long periods with no rainfall, such as southern California, Hawaii, and so forth.

Because of new construction, many heavily traveled highways are located near powerlines and substations. During the winter, a combination of snow, ice, rock salt, rust inhibitors and calcium chloride can create situations leading to bushing problems.* Other substations adjacent to railroad tracks used by oil burning locomotives are also subject to similar problems. Another environmentally critical area is near cooling towers, which produce a fine water mist that could blow into nearby lines and substations. The mist absorbs conducting oxides en route to the porcelains. In each case, the contaminants are different, but the results, tracking and failure, are the same.

Theory of Contamination Flashover

Over the past seventy years, researchers have conducted studies to investigate physical processes involved in the movement of a discharge over a moist surface. Since a completely satisfactory theory of contamination flashover has not yet been developed, various theories remain. The IEEE Working Group described flashover as follows:

> "As current flows through the conducting surface film, heat is dissipated evaporating some of the moisture, causing dry bands to form... Because these dry bands have much greater resistance than the wet surface, the voltage stress concentrates across them, occasionally leading to small, intermittent, scintillating electric discharges (from a few milliamperes to one ampere peak). The surface is wetted further, the discharge current increases until at some point the discharges elongate, join together to bridge the entire insulator and trigger a power arc. Because the flashover discharges grow along the insulator surface, the hot power arc which follows their path may damage the insulator."

---

*Snow will absorb excessive amounts of pollutants, such as salts, carbon dioxide and carbon monoxide. Then, after melting, a heavy film is left on porcelains, which can lead to run away voltage leakage. Murphy's Law activates and porcelains explode under severe electrical stress.

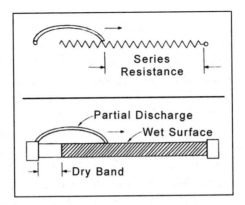

Figure 6.3 - Schematic representation of a discharge on a rod-type insulator.

Alternative Theoretical Approaches

Other proponents have also shown that their theories agree with the data from flashover tests. Hampton, for example, based his flashover theory on an experiment in which he used a water jet to simulate a contaminated long-rod insulator. Results of his study led him to conclude that flashover is a stability problem, which occurs when one of the electrical discharges crosses the dry bands and extends across the wet part of an insulator surface. Nacke and Wilkins also believe that contamination flashover is an electrical stability problem.

Although most of these flashover theories seem to suggest that flashover mechanism is a stability problem, Jolly believed that flashover is primarily an electrical breakdown process caused by the field concentration at the discharge tip. Such a process occurs in the following series of steps.

The Flashover Process

P.J. Lambeth has said that flashover of an insulator occurs when most of the surface is covered with a wet contaminant layer of low resistivity; and often takes place under any of the following conditions:

- When an insulator is being wetted after it has been energized at normal working voltage for a prolonged period of time.

- When a wet, contaminated insulator has just been energized after its normal working voltage.

- When a wet, contaminated insulator is subjected to a transient voltage.

When any of the above conditions are present, then flashover takes place. In brief, the flashover process follows a sequence (Figure 6.4) of four stages.

1. Contamination - Low Level Ionization

    As impurities are deposited on the insulator surface, a low, shortened resistance path is created through which an electrical current can flow. A high-voltage stress occurs at these points, leading to the breakdown of the air dielectric.

2. Water Filming State

    During hot, dry weather, impurities bake on the porcelain surface due to magnetic field and heat generated by electricity and solar affluency. Rain merely moistens the hardened impurities and tracking begins. The insulating bushing eventually becomes a conductor.

3. Severe Ionization

    As the voltage stress at points of contamination increases, the phenomenon of severe ionization occurs. This condition makes its presence known in the form of rays, streamers, or fingers.

4. Flashover

   The ionization phenomenon merges into the electrical over-stress type of glow discharge called corona.* Conductivity increases and at some critical voltage stress, arcing-over or flashover occurs. Flashover could also occur between phases (Figure 6.5).

   Flashover is only one aspect of bushing insulator problems. Other undesirable effects are known to exist.

   It should be noted that some of the flashover problems have been remedied by the use of bushings with resistive glaze. The glaze on the porelain has resistive material in it to where a small current is always passing over the outside of the bushing. This eliminates the possibility of forming a potential difference across a dry band on the porcelain leading to a flashover. However, this may result in high C_H and C_1 power factors in transformers equipped with these bushings due to the highly resistive surface leakage.

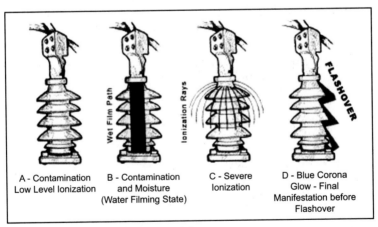

A - Contamination Low Level Ionization
B - Contamination and Moisture (Water Filming State)
C - Severe Ionization
D - Blue Corona Glow - Final Manifestation before Flashover

Figure 6.4 - The flashover process: A four stage sequence.

* Corona for transmission lines and power stations is defined as a "luminous discharge due to ionization at the air surrounding a conductor around which exists a voltage gradient exceeding a certain critical value."

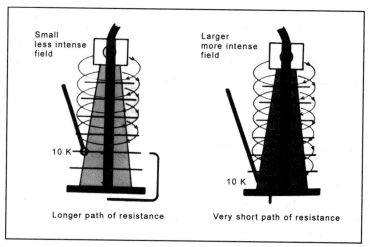

Figure 6.5 - Potential flashover between transformer bushings in different phases. If two phases were heavily contaminated (conductants) the potential for phase-to-phase flashover exists.

Other Undesirable Effects of Insulator Contamination

Insulator contamination can be detected by our external senses. Such effects include corona manifestations, and corrosion of metalwork. Thus, it is clear that few bushing failures occur without some warning.

Corona Manifestations

Four types of corona manifestations are generally associated with electrical overstress.* "Visual-corona" is perhaps the most common one, and can be observed in the predawn darkness (or at dawn) as blue-violet light dancing (so-called scintillation activity) over insulator surfaces (corona glow).

The second type is "audible corona" or "audible frying" which sounds like running water or frying grease. A smell of ozone indicates discharges are present. Finally, and perhaps the most serious type of corona discharges is the electrical effect which causes

---

* "Pre-corona" under a ultra-steep wave immediately precedes steep wave flashover.

"radio" (RI) and "television interference." It often results from field distortion caused, for instance, by an incomplete wet-contamination layer. When such signs occur, bushings should be checked more frequently because flashover is imminent, and power shut-down not far behind.

Corrosion of Metalwork

Contamination causing flashover is also likely to influence the metalwork. High leakage currents may cause corrosion through the process of electrolysis. This phenomenon seems to be significant only for DC insulators.

The above undesirable effects of corrosion can be checked by on-site periodic maintenance, which can consist of infrared inspection or through traditional visual inspection while insulator structures are de-energized, and power factor testing. Areas to look for during visual inspection include: cracked, broken or chipped porcelain, deterioration of cemented joints, gasket on solder seal leaks, moisture, evidence of arcing, broken mechanical connections, discolored oil, or oil leaks (in oil-filled units), and contamination build-up on the porcelain.

Visual Inspection

Insulator problems can be detected in two ways - by visual inspection and by infrared inspection (IR). Items requiring attention on a visual inspection include the following:

1. Porcelain - Porcelain can be chipped, cracked, or broken. Glass finish can be roughened or eroded. Chips, if small, are of minor importance and can be painted with Glyptal paint or the like to obtain a glossy finish. Minor cracks can be sealed with paint or epoxy provided the mechanical strength of the unit has not been impaired. If this is the case, the insulator must be replaced.

2. Metal Parts - These can be checked for corrosion, and painted as necessary.

3. Gaskets - Gasket materials will deteriorate with age, depending on the temperature surrounding the gasket surface and the time at that temperature. In most cases, the physical properties of gaskets will change progressively if high temperatures are maintained for long periods. Such changes could result in the reduction of dielectric strength through loss of gasket seal. Most gasket materials are badly affected by ozone which is present with corona. Therefore, it is important to inspect gaskets because they can become major sources of leaks. Care must be taken to obtain the right material and the proper thickness (for compression).

4. Capacitance Taps - These should be examined for proper gasketing to prevent the entrance of moisture. The tap compartment should be filled with an insulating compound (oil, petroleum). The ground should be inspected on grounded taps.

5. Oil Level - The oil level should be checked by inspecting the level indicators. If there is a leak, it should be repaired, and oil added. If there is no sign of a leak but the oil level is low, the problem may be internal, and it may be more prudent to replace the bushing.

6. Cement - Cement is used to attach metal parts to porcelain. It may also be used to join insulators together. Crumbing or chipping cement can be repaired in the field.

7. Lower End - The oil level in the apparatus should be checked to ensure that it covers the bushing end to the proper operating point.

8. Contamination - The bushings and standoff insulators are subject to the same contamination as the transformer and steel paints. Build-up of contamination can cause more problems with porcelain than with metal parts. These contaminants should, therefore, be removed.

Preventative Maintenance

Several practical methods of taking the offensive in bushing/insulator maintenance are available, and have been used with varying success. These methods help keep conventional porcelain and glass insulator surfaces clean, and thus prevent contamination flashover.

Periodic bushing and insulator maintenance can be achieved through one or more of the following methods:

- Hand wiping when the units are de-energized.

- Pressure washing when units are either energized or de-energized.

- Dry cleaning with air-blast materials, when units are either energized or de-energized.

- Applying protective coatings, either energized or de-energized.

Overinsulation can also be considered as another technique for preventing bushing and insulator contamination.

Hand Wiping

Hand cleaning involves cleaning insulator surfaces covered with dust, and other surface pollutants. The practice is normally used in areas where insulators cannot be cleaned by high-pressure washing or are inaccessible, or are too close to energized equipment. Although hand wiping is an efficient method of cleaning insulators, it is tedious, time consuming, and expensive in that equipment has to be de-energized.

Materials used for hand wiping may include dry wiping rags for soft, loose deposits, and wet or kerosene-soaked cloth, solvents, chemicals, and brushes or steel wool for insulator surfaces covered with caked deposits.

Because hand wiping has to be done on de-energized bushings and insulators, there is always the problem of downtime. In addition, two factors beyond the price of premium time labor are: workers are exposed to the possibility of falling, and cleanliness of the equipment is in direct proportion to the fatigue of the workers. Cement deposits also require the use of tools to chip and scrape the coating loose which can naturally cause damage to the surface being cleaned.

Periodic Insulator Washing

Cleaning bushings and insulators with a high-pressure water stream has been found to be an effective technique of removing only water reducible, non-adherent films from insulator surfaces. The method is often used in areas where natural washing by rain is inadequate to clean insulator contamination, such as California and other coastal areas. The frequency with which insulators should be washed depends on the environment in which the structures are located. Nonetheless, it is important that insulators be washed before they reach a critical level of contamination. As a general rule, such a level can be estimated from past experience on periods between flashovers, from results of equivalent salt-deposit density (ESDD) measurement, and the presence of corrosion and corona manifestations.

1. Washing equipment

    Most washing equipment in current use has been designed by the equipment manufacturer, utility companies, or by both groups. Such equipment includes different types of nozzles, hoses, and trucks.

Three major types of nozzles used for insulator washing are:

- Portable, hand-type sprayer
- Fixed-spray nozzle
- Remote-control jet nozzle

The first can be easily carried around, and is used for either regular maintenance or for emergency washing. Portable nozzles, which consist of a stainless tip and a brass section, are usually designed for individual use. In most transmission line washing jobs, nozzles are usually designed for specific spray patterns and can be obtained commercially.

The type of hose used for insulator washing is usually 3/4 or one inch in diameter, with a minimum bursting pressure of 3200 psi. When using the hose, it is advisable to pull it off completely from the reel so as to maintain the expected pressure.

Tank trucks are also a major aspect of insulator washing equipment. Two types are often employed - the portable nozzle washer truck, which has no boom and the remote control nozzle washer truck. The latter vehicle is usually equipped with a boom that is hydraulically operated, and makes it possible to position the nozzle within desired washing distance from the insulator structures. Both types of trucks are often equipped with 300- to 2900- gallon water tanks. The tanks are usually made of stainless steel, aluminum, or fiberglass material.

2. Portable Hot-Line Washing Equipment

Hot-line washing equipment has proven to be an effective and economical technique for removing pollutants from energized insulators. Hot-line washing of bushings and insulators should not simply be considered as a mere washing practice. Washing cannot be done in freezing weather, and the water must be continually tested to make sure that it is free of minerals, and has a resistivity of 150,000 ohms or better.

Live-line washing should be performed by qualified personnel to avoid serious flashover faults.

When conducting hot-line washing, it is important that the established procedures and parameters be strictly adhered to. These include:

- Verification of safety criteria by each company.

- Provision of a safe level of leakage current in the wash water stream. Leakage current is often influenced by water parameters such as nozzle-conductor distance, water resistivity, water pressure, and nozzle orifice diameter.

- Provision of maximum safety for personnel involved in the washing process in case a flashover takes place.

It should be noted the method is fairly safe. In 1937, an investigation was made into fire fighting on energized systems. This revealed that by using a diffused stream (hollow core) of water, the leakage current was quite small (0.12 milliamps). One company successfully tested their spray nozzles by spraying water on hardware cloth energized to 70,000 volts to ground.

3. Washing the Insulators

One method used for hot-line washing is fixed spray washing, in which low pressures (50 to 100 psi) are employed in areas where regular washing is required.

The method is effective in preventing sea salt contamination flashover faults. In employing this method, local parameters that influence washing, and contamination severity should be taken into consideration.

Another method involves the washing of insulator strings in which a stream of water is directed first at the insulator nearest the energized conductor, so as to take advantage of both the impact and the swirling action of the water to remove contamination deposits. After washing the line insulators in the string, the wash stream is then moved directly back on the clean units below to re-rinse them. The procedure is repeated until all the units in the string are thoroughly cleaned.

Live-line, high pressure washing at 60 day intervals is effective only in removing water soluble and non-adhering contaminants, but not in soluble chemicals such as hardened salt or cement dust. Additional cleaning of protective devices will probably be required. In fact, hard crusts under live water washing may cause flashover.

Dry Air-Blasting With Non-Abrasive Material

Powder blasting of energized or de-energized transformers is another very effective and economical technique of cleaning insulators. It does not damage the equipment; yet it will remove all insoluble particulate contamination - including metallic oxides, cement dust, chemicals, salt-cake, acid, smog, and other pollutants. One of the most common abrasive materials used for removing old crusted silicone grease and other soft pollutants from insulators is ground-up corn-cob. The other common medium is limestone powder with a Rockwell hardness of number 4, which is softer than porcelain glaze number 6. Since limestone powder is a non-conductive material, the possibility of flashover during energized work is practically impossible, except during rainfall (Figure 6.6).

A light dust film left after cleaning will wash away at the first rain shower, leaving absolutely no residue. Limestone powder is not harmful as it cleans and polished the bushing insulator surfaces to a factory-fresh finish. Other materials such as potter's clay, crushed coconut shell or ground walnut or pecan shells are sometimes used for removing hard deposits such as compacted cement dust.

If properly performed, powder blast cleaning of bushings and insulators can be safely performed up to 161 kV, using a bucket truck or a cherry picker with an insulated boom (Figure 6.7). Such a truck is usually equipped with air pumps capable of carrying air pressure of about 150 psi, a storage tank in which the abrasive materials are held, and a reel for holding the hose. Powder blasting can also be performed from other aerial locations such as substation structures or off-towers. Some applications may require the use of a helicopter to successfully reach the insulators on large electrical towers.

Consumers of electrical power have traditionally used two methods to protect bushings and insulators from contamination flashovers: coating the insulator structures with silicone or other compounds, and overinsulation.

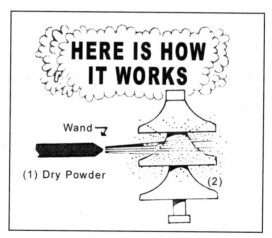

Figure 6.6 - How dry powder blasting works. Dry powder (1) under air pressure blasts the ceramic insulator, and (2) this combination penetrates the contaminant, rebounding from the insulator surface carrying the contaminant with it.

Figure 6.7 - Powder blast cleaning of transformer bushings from a bucket truck.

Use of Dielectric Compounds

Grease compounds applied on insulator surfaces serve two major functions. First they tend to "encapsulate" conducting contaminants and render them harmless. When contaminating particles land on a grease-coated surface, the grease compound fluid by osmosis 'reaches out' and engulfs the particle. This prevents a path of contamination. Second, the dielectric compound prevents water-filming and leakage currents over extended periods under wet service conditions. One product in use today, Sil-Gard®, contains approximately 80-90% silicone. The silicone causes water to bead-up on its surface, thus preventing a conductive film from forming (its hydrophobic characteristic). Figure 6.8 illustrates the effect of silicone on contaminating particles.

It is for these two reasons that silicone compounds have been extensively used (since 1953) for coating insulator surfaces. In the United States, silicone compounds are generally used, whereas petrolatums (petroleum jellies) are extensively used in Europe, and on a limited scale in the United States.

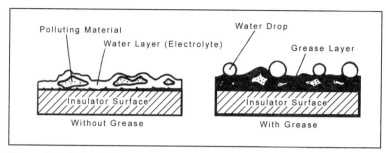

Figure 6.8 - Effect of silicone compound on insulator surface. The only function that silicone basically performs is preserving insulator dielectric surface integrity by preventing moisture from reaching engulfed pollutants. Thus, silicone is a water-repellent compound.

Conventional silicone compounds are usually made from a mixture of silica filler and a silicone fluid - an active component which provides surface mobility and water repellency. Silicones maintain virtually the same viscosity, and therefore, a temperature range of -50 °C to 200 °C.

High silicone effectiveness requires that insulator surfaces be thoroughly cleaned and dried. Insufficient cleaning will result in tracking underneath the grease compound. This in turn, can lead to permanent damage to the surface glaze and under severe stress, to cracking, breaking, chipping, or insulator disintegration.

Often an insulator surface has been thoroughly cleaned. New silicone compound should be applied evenly over the entire insulator surface to a thickness of 1/16 to 3/16 of an inch. The thickness of the silicone coating will generally depend on the nature of the contaminants and the geographical location of the substation. Where petroleum jellies are applied on insulator surfaces, the coating should be much thicker than that of silicone compounds. A major disadvantage of using a thinner layer of compound is that the surface of the insulator might be damaged when sparking takes place. This often results in the cracking of the bushings and insulators; and the eroding of glazed porcelain surfaces; toughened-glass insulators are especially vulnerable. It is for this reason that silicone compound is not recommended for use on glass insulators.

There are several problems associated with using silicone on bushings and standoff insulators. The compound is very expensive. Once the silicone has been applied, it is difficult to remove, and can only be be removed by a rag soaked in a solvent, or by corn-cob blasting (energized) and in some cases, even this is totally ineffective. Water washing is even less effective. In fact, if the silicone is not removed when "used up," the problem can be worse than if no compound has been used at all.

Another problem is how to tell when the silicone has reached the end of its usefulness. Unfortunately, the "best" method to date is the "finger test." This is done by wiping one's finger across the silicone surface (de-energized) to see two things: how clean the subsurface is, and how hard the coating has become. Silicone must be applied to a clean surface. When it is reapplied, it must also be put on a clean surface. Care must be taken in applying the compound as it is very slippery, and should only be put on insulators and insulator caps and slightly on metal heads of oil filled bushings, not steel.

The major advantage of silicone compound in protecting against contamination buildup is this: if an insulator protected with the compound flashes over, the porcelain glaze is damaged less than if it were unprotected. The arc by- product of silicone is silica, which is non-conductive, whereas, the arc by-product of petroleum jelly is carbon, which can be conductive.*

In addition, the bushing surfaces should be clean before a valid insulator power factor test is made. While a slow increase of power factor may be normal, degradation of that portion of the insulation which operates at a greatly elevated temperature will result in a substantial increase in power factor. An unusual increase in power factor may become an indicator of the detrimental effects of loading beyond nameplate rating. Bushings which have been loaded beyond nameplate rating should be measured more frequently. Since power factor varies with temperature, all measurements should be made at 20 °C to 25 °C or corrected to the reference temperature.

*Pure carbon is a good resistor: contaminate carbon is not.

Grease compounds can be applied to de-energized insulators by hand, using rubber gloves, pads or brushes, or by air-spray equipment. In order to apply the grease compound more uniformly, the air spray method is generally recommended.

Where insulators are energized, "hot stick" air spray equipment is generally used, and should meet all the state and OSHA requirements for hot-line maintenance.

The effective service-life of silicone grease can vary from one month to five years, depending on the thickness and evenness of silicone layer applied, and the geographical location in which the insulators are situated.* For severely polluted environments, one major utility has reported an annual or semi-annual scheduling for use of silicone compound. Another facility, located in a nonpolluted area, has gone to no more than two years between energized cleaning.

It should thus be appreciated that even with absolutely effective power blast cleaning, grease compounds will gradually lose their water repellency and arc resistance due to extended exposure to high-voltage corona, ultra-violet light, water erosion, and/or particulate contamination. When this condition occurs, grease compounds should be removed by energized blasting using corn-cob.**

Overinsulation

Overinsulation as a method of checking contamination flashover is advantageous in that only infrequent maintenance is required. In addition, the bushings and insulators are standard items which can be readily obtained from manufacturers or from suppliers. Thus, overinsulation, consisting of additional stacks of insulators, at best only moderately decreases maintenance cost. Since airborne particles will cover a large insulator at the same rate

---

*The evenness of the film is as critical as its thickness.
**A bushing will dry silicone compound at a faster rate than an insulator due to rising heat and elongated magnetic field. (Figure 6.2)

Typical Quantities of Silicone Grease Required to Coat a Specific Insulator[1]

Insulator Type and Description[2]	Area (Square Feet)	Amount of Compound Required	
		Pounds	Ounces
13 kV one-piece pintype	0.70	0	3.6
34.5 kV line-post insulator	2.32	0	12.1
Standard 10-inch suspension unit	1.82	0	9.5
Smogtype 10-inch suspension unit	2.47	0	12.5
34.5 kV pin-cap apparatus insulator	4.24	1	6.0
69 kV pin-cap unit used in two unit stacks	4.96	1	10.0
High voltage pin-cap apparatus unit	7.73	2	9.0
69 kV bushing	9.1	3	-
138 kV bushing	21.4	7	-
230 kV bushing	42.4	16	-

[1] If a given insulator were coated with a 1/16 inch coat of compound, each insulator would require the amount of compound listed.
[2] Courtesy of Ohio Brass Co., Mansfield, OH.

Table 6.2

as stacked small ones, this method is not as effective as reasoning would indicate.

Increasing insulator size or creepage length has proved not to be the ultimate answer to flashover problems in severely contaminated areas. Nonetheless, vertical mounting of insulators - in contrast to horizontal - has proven beneficial in minimizing contamination problems. Again, common sense and experience reminds us that preventative maintenance is the ultimate solution to clean bushings and insulators.

The Last Word on Cleaning

Each substation presents a unique problem dependent upon many factors such as the length of dry season, fog, salt spray, industrial contaminants, design, and prevailing winds. Clients have recognized that regularly scheduled powder-blast cleaning is the best approach in eliminating unscheduled power outages. They also recognize that a high voltage insulating coating is an integral part of their bushing-insulator maintenance system in 40% of all cleaning situations, regardless of the environment. Silicone has been most often chosen to combat environments of salt-cake and cement dust.

Nevertheless, the use of coatings is not beneficial in every situation. For example, one substation located near a cooling tower should be treated, while another substation located only 1,000 feet further away would not require treatment. The primary parameter for usage is actually simple. If moisture is your problem (historically), water-repellent silicone is your likely answer.

The decision on the type of maintenance program and its frequency depends mainly on the effluency of airborne contamination. Criteria that will also assist in making a decision include construction type, condition and age of bushings and insulators in-service.

Keep in mind - "Surface contamination and leakage currents caused operating difficulties eighty years ago... and exist today as one of the most serious insulator problems." Bushings and insulators still are not maintenance-free. In fact, maintenance-free insulators have been the subject of research for years. But as an IEEE conference paper has reported, "Contamination-proof insulator designs do not exist and are unlikely." At this point, the classification and tabulation of more than 40 design styles of insulators support this view. Even in light of new insulator designs - light-weight rubber, epoxies, synthetic resins, and glass-flake - bushing and insulator contamination, will likely remain a problem plant maintenance personnel cannot avoid." Additionial information can be found in the IEEE Std 957-1995, *IEEE Guide for Insulator Cleaning*.

Internal Failure

An internal failure of a bushing can manifest itself from different causes.

- Moisture ingress due to hairline cracks in the porcelain housing.

- Moisture ingress due to a faulty seal of the bushing top particularly the flex seal design from General Electric.

- Moisture ingress due to dissimilar metals (for example, aluminum and copper on some bushings cause heating).

- Failure of the condenser due to partial discharge, tracking, migration of conducting ink, and wrinkled foil on metal condensers.

- Overheating of the bushing leading to overpressure of the gas space. Rapid cooling may produce a gas bubble from the gas-saturated oil. The tendency for this to happen is in the area of the highest electrical stress. This will lead to partial discharge. Overheating may also be caused by the use of dissimilar metals in the construction of the bushing.

Internal Failure Detection

- Testing the bushings with the bushing tap test, or sometimes refered to as the power-factor tap, will give an indication of the internal condition of the bushing. Modern condenser type bushings rated 15 kV - 69 kV are usually equipped with a test tap. This will allow testing of the bushing without removing the bushing from the transformer.

- Bushings rated above 69 kV are usually equipped with a potential tap (also known as a capacitance tap). This also allows bushing testing without removal from the unit.

- Using a power factor test set, measurements are made of the C_1, (insulation between the center conductor and the tap), and C_2, (the tap to ground insulation). The results are then compared to nameplate values or previous benchmarked values.

- Changes in the power factor or capacitance valves may be an indication of high resistance, open circuits, moisture, tracking, overheating, loss of ground, ink migration, and problems with the insulating fluid in the bushing. Problems like these typically do not heal themselves, so great care should be taken when a change is noted.

Chapter 7

Transformer Life Extension

Introduction

When oil testing indicates that a problem of moisture, certain dissolved gas and/or decay products such as acids and sludge exists, corrective maintenance procedures need to be followed. These procedures must be followed if you intend to maintain the reliable operating life of the transformer.

Water trapped in oil-impregnated cellulosic materials, indicated by Karl Fischer moisture analysis in terms of a high percent of moisture saturation or a high-percent moisture by dry weight and confirmed by winding power factor and other electrical tests, must be removed through some type of dehydration of the transformer insulation system. Dehydration methods can only reduce the rate of loss of mechanical strength caused by the moisture contamination. Mechanical strength lost due to moisture contamination cannot be regained. The destructive process can only be retarded.

Gas in oil analysis gives a profile of dissolved gases, including combustible gases present in the in-service transformer oil. The removal of these gases may require some type of repair to be done on the transformer to cease the formation of the gases in question. The oil should be degassed after the repairs have been made (just removing gas will not fix the problem that caused it). This allows the transformer operator to establish a new baseline of dissolved gases for the transformer.

Oil decay products that are deposited in the form of sludge in and on vital transformer parts will require some method for the removal of the sludge. We will consider how these three groups of impurities - moisture, dissolved gas and sludge, can all be effectively removed through one process, Hot Oil Cleaning®. This process can be done while the transformer is on-line and operating, with few exceptions.

What To Do If You Find Free Water In a Transformer

When routine sampling finds free water (water visible to the naked eye, separated from oil) in a transformer, certain steps need to be taken immediately. (Figure 7.0)

First, determine how a quantity of water like that got into the transformer. Perhaps there is a cracked bushing, bad gasket or some other type of leak. If there is a vacuum on the headspace of the transformer, this can compound the problem of moisture ingress. Perhaps an inspection cover is not properly installed. There can be a number of different ways which quantities of water can get into a transformer. Find out the reason affecting yours.

Second, determine through oil testing whether or not the transformer can be dehydrated while it is energized. Karl Fischer moisture parts per million, percent saturation and percent moisture by dry weight will determine if dehydration can happen while the transformer is energized.

Third, if laboratory tests on the oil indicate that the transformer should not be dehydrated while the transformer is energized, de-energize the transformer and remove the connections from the transformer's primary and secondary loads at the bushings. After the connections have been removed, electrically test the transformer to determine the extent of moisture contamination.

You don't necessarily need to panic if free water appears in your transformer. It depends on how the water entered the transformer.

If free water is discovered when attempting to draw an oil sample from a transformer, follow these steps:

Figure 7.0 - Free water and oil

1. Apply positive pressure to the headspace of the transformer with dry nitrogen that meets *ASTM D 1933 - 97 Type III* specifications. Do not exceed three pounds per square inch (psi) or 20 kPA.

2. Determine where the leak is by applying a soap solution or commercially available leak detection solution around all joints and fittings. Leak detection systems that utilize helium gas and helium sensors may also be used.

3. Make the necessary repairs and check again for leaks. (see Transformer Leaks and Leak Repair, pg 266)

4. Inspect the inside of the transformer tank and tap changer compartment to determine the location of free water leaks. Water could be entering the system from a variety of different sources, bad gaskets, cracked insulators, rust holes, loose inspection covers, and ruptured explosion diaphragms.

These steps should help determine how this moisture contamination got into the transformer insulation system. The next thing to determine is whether or not the windings are wet and to what extent they may be wet.

Remember that cellulose in the windings is very hygroscopic (absorbs and adsorbs water), especially at lower temperatures. Transformer oil is hydrophobic (literally, afraid of water) and will tend to give up dissolved water, where it is attracted to the cellulose insulation. Cellulose insulation has 300 to 3000 times the affinity for water as does oil - which varies as a function of temperature.

Large amounts of free water entering a transformer may or may not affect the windings, depending on the location of the leak. If the water enters the transformer from the top of the unit, it will probably go on and over the windings and be absorbed into the cellulose insulation. If the water enters from the side, it may simply run down the inside of the transformer tank and settle at the bottom of the unit.

Stories have been told of transformers with oil levels so low that windings were not insulated with oil and cooling tubes were rendered useless because low levels prevented circulation of oil. This caused the transformer to operate hotter than it should. In order to correct this situation without waiting for the delivery of transformer oil, water was very slowly and carefully pumped into the bottom filter press valve of the transformer. This would raise the oil level up so that the windings were protected and the cooling tubes were once again operating as they were designed to operate. The water was drained out and replaced with oil when the oil became available. The transformers operated fine for the rest of the normal lives. While this procedure is not recommended, it serves to prove that free water is no reason to panic about the condition of a transformer.

The applications of winding insulation tests that will be discussed are the minimum tests necessary in identifying the problem of moisture and sludge in a transformer. These tests are nondestructive tests, so there is no risk to the integrity of the insulation because of the testing. Once a leak in a transformer is repaired, it is quite easy to dry out the oil. Drying out the windings is a much more complex procedure.

Transformer Leaks and Leak Repair

Transformers develop leaks for a variety of reasons. One of the most common is the breakdown or deterioration of the gasket material. Some other sources of leaks are: damage to the tank or accessories, failed or poorly welded seems, over-filling of the dielectric fluid, damage to an internal barrier board, damage to a conservator bladder/air cell bag, faulty valves, over-heating or an over-pressure situation.

Most leaks are apparent and easily located by a wet or oily stain in the area of the leak. Sometimes the leak will be more difficult to locate and a more extensive leak test procedure will need to be performed. This would include thoroughly cleaning the suspected area; apply a positive pressure with dry nitrogen and applying a detection agent.

Once the leak is located you will have to determine if the unit is field repairable. Will it be necessary to ship it to a repair facility, or is it worth repairing at all? Some manufacturing methods make it extremely difficult for repairs to be made in the field. Some examples are: bushings that pressed into a plate and then welded onto the unit, welded flanges and units with no access covers to facilitate bushing removal. In some cases, especially with the smaller kVA sizes, it may be more cost effective to replace the unit.

Many things will determine the success of the repair, but one of the most important things is to be prepared. Have all the parts and material that you will possibly need on hand. It would be best to find exact replacement parts from the manufacture. If that is not possible, make sure the after-market parts will fit and function properly. Most after-market parts suppliers should be able to provide a comparison analysis sheet. This would be most important for bushing replacement.

The material that you select for gasket replacement is critical. Gaskets have several important jobs in sealing systems on a transformer. A gasket must create a seal and hold it over a long period of time. It must be impervious and not contaminate the insulating fluid or gas above the fluid. It should be easily removed and replaced. It must be elastic enough to flow into imperfections on the sealing surfaces. It must withstand high and low temperatures and remain resilient enough to hold the seal even with joint movement from expansion, contraction, and vibration. It must be resilient enough to not take a "set" (becoming hard and losing its resilience and elasticity) even though exposed for a long time to pressure applied with bolt torque and temperature changes. It must have sufficient strength to resist crushing under applied load and resist blowout under system pressure or vacuum. It must maintain its integrity while being handled or installed. If a gasket fails to meet any of these criteria, a leak will result. Gasket leaks result from improper torque, improper sealing surface preparation, choosing the wrong type gasket material, or the wrong size gasket. Gasket thickness is determined by groove depth and standard gasket thickness. Choose the sheet thickness so that one-fourth to one-third of the gasket will protrude above the groove; this is the amount available to be compressed. Gasket sheets come in standard thickness and increase at 1/16-inch increments. Choose one that allows one-third of the gasket to stick out above the groove if you can, but never choose a thickness that allows less than one-fourth or as much as one-half to protrude above the groove.

Neoprene, Cork-Neoprene, Nitrile (Buna N), Cork-Nitrile, and Viton all have different applications. Consult the gasket manufacturer and make sure the material is compatible with the minimum and maximum operating temperatures, type of fluid, internal pressure and type of joint the material will be used in.

After the unit is repaired the unit should be leak tested to assure that it has a tight seal. Failing to do so may result in a leak in the near future that will require additional repairs costing valuable time and money.

Dry nitrogen is applied to the unit. Depending on the pressure rating and the size of the unit, this could range from 3 to 8 psi, and recommend time could range from 1 to 24 hours. The area can then be checked with a soapy water substance and or with a pressure gauge attached to the unit. Depending on the size and kVA rating of the unit, a specific sit time so that any air bubbles have time to leach out, and dew point measurements to determine if any moisture was introduced to the unit may be required.

After it has been determined the unit has been repaired properly and has no moisture problems, it can be re-filled. If applicable, follow manufactures recommended fill procedures.

Electrical Tests to Determine Moisture Levels in Solid Insulation

The Power Factor Series

A very effective test to determine the moisture level of solid insulation is a power factor test, also referred to as the Doble Test. It is an AC Test and extremely sensitive to moisture. Depending on the voltage class and MVA rating of the transformer under test, power factors as high as 1% may be acceptable. In other situations, power factor readings in excess of 0.5% may be unacceptable.

A good way to determine if moisture is the cause of the unacceptable power factor reading is to do the tip-up (or tip-down) test. If we have a test result that is considered to be high at the 10 kV test voltage, we will do the test again at a lower voltage. If we see a decrease in the power factor reading as the test voltage is decreased, the problem is voltage sensitive, and therefore we would not believe it is a moisture problem. If the power factor virtually remained the same, moisture would be the reason for the high power factor.* (see Table 7.1)

The Megohm Meter Series

The megohm meter series consists of four tests, the primary test being the minimum insulation resistance test at one minute. Continuing this test for a full ten minutes results in the dielectric absorption test. The dielectric absorption test is a refinement of the standard insulation resistance test. Temperature is not a factor in

Voltage Test (kV)	Percent Power Factor (C_H)
10	1.40
5	0.60
2	0.25

Table 7.1

* Use proper temperature correction tables when testing unit at temperatures other than 20 °C.

the dielectric absorption test and conclusive data can be established without consulting previous test records.

Using a motor driven megohm meter, dc voltage is applied to the equipment being tested for a period of time, usually ten minutes. Insulation resistance readings are recorded at specific times (15, 30 and 45 seconds, 1 minute and each successive minute up to the ten-minute test time conclusion). The results are plotted on log-log paper. Insulation in good condition will have a dielectric absorption curve that is a straight line, increasing with testing time. Insulation with moisture or contamination will result in a line that rises slowly, if at all and flattens out over time. The temperature of the unit will have a direct effect on the results. Use the proper temperature correction tables when testing at temperatures other than 20 °C.

A polarization index of less than 1.0 may indicate excessive moisture contamination. If the polarization index ratio (ten minutes divided by the one-minute reading) is less than 1.0, moisture may have been absorbed into the windings. The oil and the windings will need to be dehydrated to improve this ratio. The polarization index for a good dry transformer should be in the 2.0 range.

Another test that can reconfirm moisture contamination is the step-voltage test using the megohm meter. If the test displays a lower reading in megohms at the higher applied voltage, the insulation is contaminated. Weak insulation will have a lower resistance at the higher voltage.

The insulation tests just mentioned, power factor, polarization index dielectric absorption, tip up, minimum insulation resistance and step voltage, are beneficial tests for determining whether or not transformer windings are wet. See Chapter 4 - Electrical Testing for description of tests.

Dehydration of Transformers

Introduction

The continued reliability of a transformer's insulation system requires the highest level of dryness in that system. It is especially important for high-voltage transformers (115 kV primary and higher), if for no other reason than to protect the million-dollar investment involved with high voltage transformers. All drying methods share a common objective, the removal of free and dissolved moisture from the cellulose insulation as well as the insulating fluid.

When do you need to dehydrate your transformers? There are two motivations for transformer dehydration. One is to avoid near-term failure. The other is to extend (or maximize) reliable transformer operating life.

Dehydration to Avoid Failure

Solid insulation has a much greater affinity for moisture than does liquid insulation. Depending on temperature, cellulose insulation will hold from 300 to 3000 times more moisture than will oil. A new transformer should have no more than 0.5% moisture by dry weight (%M/DW) if it has been properly dried. An operating transformer that has been in service for a number of years should have no more than 1.0% M/DW.

Gradual increases in the percent moisture by dry weight of the solid insulation may deduct years off a transformer's life but pose no threat of immediate failure. Not until solid insulation approaches 4.5% M/DW does the moisture content pose a danger of immediate, electrical failure. At or above 4.5% M/DW a transformer runs a very real risk of flashover across the insulation at normal and elevated operating temperatures. The moisture absorbed in the insulation will form a conductive path from winding to winding, to the core or to a ground.

Dehydration to Extend Transformer Life

It may not be practical to maintain every transformer in your system at or below 0.5% M/DW. You may choose to let the moisture level in distribution class transformers to elevate to the 1.0% M/DW range. But, additional moisture in the solid insulation beyond 0.5% M/DW decreases transformer life. In order to optimize transformer life, you should maintain your insulation less than 0.5% M/DW.

Factors concerning the choice of drying method include the type of transformer, ambient and operating temperatures, time limitations, thickness of the pieces of insulation and the dew point of the moisture in the surrounding atmosphere. Drying methods utilize heat and vacuum and are done in the factory and in the field. Depending on the moisture ppm and the percent moisture by dry weight, this work can be done on an energized or de-energized basis.

Insulation Drying Practices

To illustrate the difficulty of dehydrating a transformer's solid insulation, imagine plunging a phone book into a bucket of water and leaving it there for a week. It would be easy, at the end of the week, to dump the water out of the bucket. Drying out all the inner pages of the book would take much more work. Just as it is easy to dump the water out of the bucket, it is also relatively easy to remove water from the oil in a transformer tank. Removing the water absorbed by solid insulation inside the tank, like drying out the phone book, is much more difficult.

Before any dehydration technique begins, care must be given to the possibility of static charges building up. Static charges can accumulate when transformer oil flows through hoses, pipes and tanks. Combined with air and oil vapors, there is a chance for an explosion if proper methods are not utilized. All devices should be grounded. This may eliminate any harmful discharge of explosive vapors. Along with grounding, all hoses should be bonded at their metal fittings to ensure safety.

Factory and Repair Shop Methods

There are two main ways that insulation is dried in the factory using only heat. The first method is known as oven drying. In oven drying, the core and windings are removed from the tank and placed into a large oven (see Figure 7.1).

The core is baked at high temperatures over a period of time. The cellulose insulation is dried with this method but dissolved moisture in the oil is a separate problem.

Another method of drying insulation in the factory is with elevated heat with the use of dry air. In a shop, heated air is circulated through the inside of a transformer. The exterior of the transformer is insulated in order to hold in the heat.

The moisture is removed from the circulated air on a continuous basis. This method of dehydration will take a long time. This method works for older insulation designs if the dew point is less than 0 °C or less than 25% relative humidity at 20 °C.

There are two other methods of insulation drying that occur in the factory that are more effective than the above mentioned methods. They are better than methods that use only heat because they also use vacuum. The Vapor Phase method of transformer insulation drying was developed by the General Electric Company. They coined the term Vapotherm® for the process. Almost all transformer manufacturers use some form of vapor phase dehydration to achieve good dry insulation that is required by transformer operators. The principles of heat and vacuum are still used in vapor phase dehydration but the method of heating is different.

Figure 7.1 - Transformer core in oven (shop).

Vapor phase dehydration begins by placing the core and coil of a transformer into a vacuum tank. The air in the tank is evacuated until the atmosphere in the tank is at approximately 35 mm Hg (Torr). This removes a large part of the wet air in the system before heat is introduced to the system. It also assures that the system is airtight. Special grade kerosene is heated to around 140 °C and introduced in the vacuum tank that is holding the core and coil.

The hot kerosene transfers heat to change liquid moisture to vapor. The kerosene vapor condenses on cooler surfaces and penetrates into cold interior parts of cellulose insulation. The solvent then gives up the heat of condensation to the object on which it condenses, causing a rise in temperature.

The heat of the kerosene increases the temperature of the core and coil when the hot vapors condense. With this rapid increase in temperature, the vacuum strips moisture from the core and coil.

The solvent vapor and water vapor are extracted and recovered at condensers. The solvent and moisture are separated. The moisture is removed from the system and the hot solvent vapors are then recirculated into the vapor phase process. Particulate matter that may be in the windings is also removed at this stage of the process.

Continuous vacuum treatment is then added to the processing tank. A vacuum reading of 0.5 Torr when the insulation has a temperature of 100 °C is desired. If these values are attained, moisture contents as low as 0.1% is possible. The vacuum removes moisture and condensed solvent from the insulation making for very dry cellulose. The cellulose is now ready to be impregnated with transformer oil.

Transformer oil is introduced under vacuum into the system at this time. This allows for complete saturation of the cellulose insulation with the oil. The core and coil assembly is then removed from the processing tank and placed in the transformer tank. Any moisture picked up during the installation of the core and coil is very minute and can be removed with processing once the transformer case is sealed.

After drying out at a shop, torque-clamping devices should be used. These devices are used to structurally tighten the core. Depending on the initial moisture content of the insulation, the paper insulation may shrink when dehydrated. If the system is not tightened, other problems may arise.

While this is a very good system for removing moisture from insulation, not all shops are equipped to perform this type of work. There is a high cost associated with starting up this system so it may only be available at larger repair shops and factories.

The other method for dehydrating wet transformer insulation in a repair shop involves applying a hot oil spray over the core and coil. Heat from the oil spray slowly penetrates the insulation.

Moisture is removed when vacuum is applied to the tank. This process is similar to vapor phase dehydration but is not as effective. Temperatures don't get as high as they do with vapor phase and it is very difficult for the inner most windings to be reached by the hot oil spray. In most cases, this method will be performed in the field, not in a repair shop.

Field Dehydration Methods

The goal of drying transformer insulation in the field is to try and get the insulation as dry as it could be if dehydrated at the factory. Standard C57.93-1995, *IEEE Guide for Installation of Liquid-Immersed Power Transformers*, lists four different methods for drying transformer insulation in the field. The methods are:

- Circulating hot oil
- Short-circuited windings (oil-filled), vacuum (without oil)
- High vacuum
- Hot air

Circulating Hot Oil

A heater is used that is capable of raising the oil temperature in the transformer tank up to 85 °C. This heat must be maintained for this process to be effective. To help maintain the heat, close any radiator valves. This will prevent the circulating and cooling of the hot oil. Insulate the transformer tank to improve the efficiency of the heater.

The hot oil is circulated throughout the transformer tank until the cellulose insulation is determined to be dry (by filling the tank with dry nitrogen and taking a dew-point measurement after 24 hours). This method is more effective if vacuum is used. However, due to the extended length of time involved with dehydration using this method, it is not used very often.

Short-Circuited Windings, Vacuum

This method of dehydration circulates current through a winding, which is the heat source for the dehydration process. This current source is connected to one winding and the other winding is short-circuited. The winding must not be overheated, since excessive heat may accelerate the aging of the insulation, which is what's trying to be prevented in the first place. Winding temperatures should not exceed 95 °C and oil temperature should not exceed 85 °C. After the desired temperatures have been reached in the oil and insulation, the current source is disconnected from the winding and the oil is drained from the transformer tank.

A cold trap and vacuum are also used with this method. When the tank is empty, vacuum is applied until the predetermined amount of moisture is removed. Melting the frozen water vapor from the cold trap and measuring its volume indicates how much moisture has been removed from the insulation. Repeat this process as often as necessary until the insulation is sufficiently dried.

High Vacuum

This dehydration procedure, also known as vacuum filling, requires a vacuum pump capable of pulling down to an absolute pressure of 50 microns (7 Pa) or lower. The transformer tank must be rated for a full vacuum in order for this method to be effective. Vacuum rating information is usually found on the nameplate, designated as capable of handling ±15 psi (sometimes shown as ± 14.7 psi, the actual limit of vacuum).

A cold trap may be used with this method so that the amount of moisture removed from the insulation can be measured. It will also prevent moisture accumulation in the vacuum pump, therefore it will be more efficient. Heat will improve the efficiency of this method and will also reduce the amount of time needed to achieve an acceptable result.

The removal of moisture from the cellulose insulation begins when the vapor pressure in the tank is lower than the vapor pressure of the moisture in the insulation. The transformer must not have leaks in order for this vapor pressure to be achieved. Therefore, any leaks must be repaired before attempting to dehydrate.

Once moisture is extracted from the insulation, the moisture content of the insulation can be estimated with the use of the Piper Chart (Figure 7.2). To do this you will need to determine the insulation temperature and final vacuum pressure. This dehydration process continues as long as moisture is being removed or until the moisture in the insulation gets to a predetermined level.

This method is the most common method used for dehydrating transformer insulation in the field. It is also the most cost-effective and efficient method available for use in the field.

Hot Air

If using this method to dehydrate transformer insulation, make certain to insulate the transformer tank. This will keep heat in the tank and allow the heater to work more efficiently. Keeping the tank warm will also prevent condensation of moisture in the tank.

Clean dry air is heated and passed through the coils of the transformer. This heated air draws moisture out of the cellulose. The air and moisture is then vented through the top of the transformer. The larger the transformer tank, the greater the volume of air needed to dry out the insulation. Caution: this method will have oil and air vapor that is flammable.

The overall most effective methods use a combination of heat, vacuum, cold trap, internal and external heat.

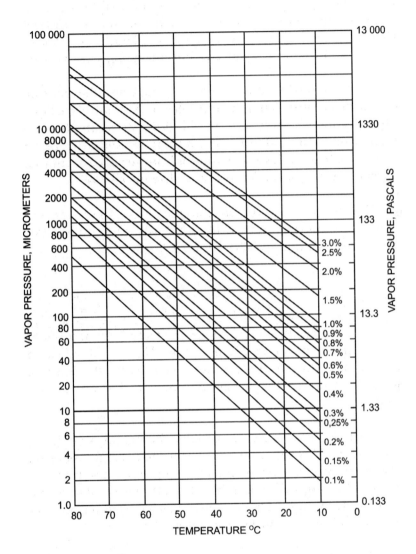

Figure 7.2 - Piper Chart

InsulDryer "On-Line" Transformer Dehydration Unit

InsulDryer is a relatively new concept in the dehydration of power and transmission transformer insulation. An InsulDryer can be the key to drying wet insulation in a transformer that cannot be taken out of service, which is a requirement for most field drying techniques. These systems are also used as a preventive maintenance tool to keep the transformers insulation dry, therefore extending the life of the transformer. Remember, "The life of the transformer is the life of the insulation."

These state of the art dehydration units are portable, easy to install and easy to service (Figure 7.3).

Figure 7.3 - InsulDryer

Results of the Drying Process

The Bonus of Degasification

A benefit to insulation dehydration that used vacuum is that the oil and paper also get degassed. This benefit does not occur in dehydration methods that do not use vacuum. Transformer operators recognize the need not only for dehydrated oil but also for degassed oil. Degassed oil minimizes partial discharge effects caused by inadequate oil impregnation, removes oxygen (the root of oxidation by-products) and removes dissolved combustible gases.

Degasification is accomplished by exposing the oil in a thin layer to a high vacuum. Coalescing filters in a vacuum chamber create what is known as a "rain forest" effect inside of the vacuum chamber. This allows for more surface area of the oil in the vacuum, which allows the vacuum to work more efficiently in dehydrating and degassing transformer oil.

Drying Efficiency

The efficiency of dehydration, regardless of method, largely depends on two factors; oil temperature and level of vacuum pulled. The challenge is to use a system that will not use too much heat, which will degrade the cellulose, yet will still remove moisture from the cellulose.

Factory residual moisture content (0.3-0.5%) shows equilibrium with vacuum as follows (info taken from Piper Chart):

- At 40 °C, minimum of 70-200 microns (10-27 Pa) vacuum required

- At 60 °C, minimum of 300-1200 microns (40-160 Pa) vacuum required

- At 80 °C, minimum of 1200-1400 microns (160-190 Pa) vacuum required

For efficient dryout, vacuums in excess of the equilibrium values are required. The drying operation occurs when the oil-cellulose equilibrium is disturbed. Other factors that determine the efficiency of water removal include the initial water content in the oil and cellulose, oil film thickness and time of exposure to vacuum.

Dehydration and Degasification Only?

In the case of only wet oil or wet oil with wet windings, a couple of circulations of the oil through a vacuum dehydration system can dry the oil. The dry oil will subsequently be contaminated by the wet windings as they seek moisture equilibrium. Continued circulations will eventually correct this situation. The circulation of the oil through an energized transformer will ultimately reduce the insulation moisture content.

EHV equipment is not the only type of transformer that needs a high degree of dryout. Most power transformers are processed in the factory to the highest degree of vacuum dryout and impregnated with oil under vacuum. Anything less than this will result in a shortened life of the transformer.

Limitations of Dehydration Systems

Today's oil purification systems are very effective in removing water from transformer oil. Eventually, the rate of water removal is so small that continued servicing becomes economically prohibitive. Although the oil is dry at this point, the process may or may not dry the transformer. This is because the rate of drying depends on the rate of diffusion of water through the paper, which initially contains over 99% of the moisture in the transformer. Therefore, after several weeks of normal operation following the initial treatment, the transformer can appear to be nearly as wet as its initial condition. The cause of this phenomenon is largely due to insulation design.

Older transformers were constructed with thick, solid insulation structures. The depth of the water in the insulation, along with the absolute amount of moisture, could require a much longer drying time. Newer transformers consist of a built-up structure in which relatively thin insulation barriers (usually less than 1/8 inch (3 mm) thick) are alternated with oil ducts of similar thickness. Nevertheless, a great deal of time is still necessary to attain the degree of dryness desired.

Transformer Oil Regeneration and Hot Oil Cleaning

Introduction

The insulating oil in electrical transformers serves four functions:

- Provide a dielectric medium
- Provide a cooling medium
- Protect the paper
- Use as a diagnostic tool to monitor the condition of the paper

The first purpose requires that the oil be free of water and suspended matter. This helps it act as a good electrical insulator. The second function requires an oil of relatively low viscosity, volatility and good heat transfer capability. Deterioration of the oil over time, through oxidation and/or contamination, can cause the transformer to overheat and prematurely fail. The third function requires oil that is clean also because the life of the transformer is the life of the paper. The fourth function gives us an indication of the effectiveness of the first three functions.

Industry and utilities can no longer afford the luxury of installing a new transformer and forgetting about it. Older transformers were over-designed and over-built. Newer transformers minimize materials and maximize stressing. Unfortunately, too many transformers have been ignored for so long that only corrective maintenance - reclamation of the oil and desludging of the transformer - can rectify the situation.

When a heavily sludged in-service oil, completely unfit for continued use in a transformer, is examined, at least 80% of the hydrocarbons present in the oil are unchanged. The oil can be reused if the oxidized products are removed. Transformer oil oxidizes due to the impurities present in the oil. Pure hydrocarbons are not easily oxidized under normal operating conditions.

With the depletion of good grade naphthenic crude oils and with changes in the refining industry, the need for a practical approach to proper use, maintenance and recycling of transformer oil is of high importance to the transformer operator. The development of new dielectric fluids and gases may help with this problem but so far hasn't been as successful as hoped. The costs of other fluids and gases are many times higher than naphthenic-based transformer oil and the properties are not necessarily better than those of oil. SF_6 gas may adversely affect the ozone layer if it escapes from the transformer. It is in the best interest of all involved with the proper operation of transformers to maintain what transformer oils are already available.

Inadequacies of Existing Standards

What rate of damage to your transformers are you willing to accept? Technical societies all around the world apparently have been willing for you to accept a great deal of damage to your transformers. If you follow the standards they have adopted you will allow your transformers to age at an avoidably accelerated rate. In the late 1990's, growing awareness has spurred on some groups to begin reviewing their standards but adequate changes have yet to be made.

Let's review some of these standards and the damage they allow:

IEEE in standard 637-1985, standard C57.106-2002 and standard 62-1995 allow an acid number up to 0.2 in transformers under 69kV, 0.15 acid in transformers up to 230 kV and 0.1 in transformers 230 kV and up.

IEC 60422-1989 and BS 5730 both allow up to 0.5 acid in operating transformers! As if that weren't bad enough, both of these standards allow replacement of the oil beyond these limits as a suitable maintenance action!

From an ASTM survey completed in 1957 we know that over 70% of transformers with oil that reaches 0.5 acid show sludge deposits visible through a top inspection plate. We can see from electron microphotographs that sludge begins accumulating in insulation paper fibers around 0.1 acid (see p. 317). These conditions must be prevented, not allowed, in order to achieve the maximum, reliable operating life from your equipment.

Insurance company standards don't help transformer owners much more than these technical standards do - especially since they usually don't take more conservative positions than the standards. An insurance company typically is concerned with this question: "Will this transformer fail this year?" If testing doesn't indicate imminent failure they will be happy to include that unit in the year's insurance coverage. When a problem does appear that could lead to failure they will (very reasonably) ask the owner to fix the problem or they will exclude it from coverage. An insurance company, though, will not insist that the owner address conditions that are slowly degrading the integrity of a transformer's insulation system. That is not their job.

Insurance companies' and technical societies' interests align with the owner's in some but not all areas. The transformer owner can accept guidance from both groups but can't entirely rely on them for all aspects of operating and maintaining transformers. No other group has yet really embraced all the concepts presented in this publication for maximizing reliable transformer life. You will need to choose them for yourself on the basis of the evidence presented here (Figure 7.4).

The Correct Oil Reclaiming (Regeneration) Strategy

Ignoring water contamination for the moment, there are essentially four categories of conditions you could find your transformers in. Working backwards from the stages of oil decay, they are:

1. Oxidized oil, visible sludge deposits.
2. Oxidized oil, no visible sludge deposits.
3. Depleted inhibitor, no oil oxidation yet detected.
4. Adequate inhibitor content.

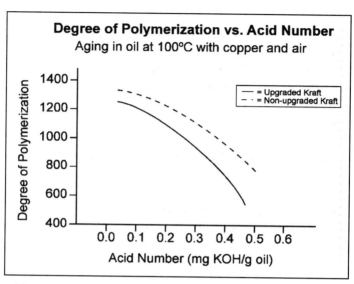

Figure 7.4 - Degree of Polymerization versus Acid Number

Here are the correct oil regeneration strategies for each condition:

Oxidized Oil, No Visible Sludge Deposits

SludgPurg® or Hot Oil Cleaning of the transformer to remove oxidation by-products from the oil and to re-dissolve and remove sludge deposits from surfaces inside the transformer.

Hot Oil Clean® to remove oxidation by-products from the oil and to clean oxidation deposits absorbed by the cellulose insulation.

Depleted Inhibitor, No Oil Oxidation Yet Detected

ReInhibit® service. This involves first cleaning oxidized inhibitor compounds from the oil before adding inhibitor that has been dissolved in clean, dry oil.

Unfortunately, just adding more inhibitor to oil with depleted inhibitor is not an effective strategy. The compounds formed by depleted oxidation inhibitors undermine new inhibitor introduced into the oil. Figure 7.5 shows the relative life of clean, inhibited oil compared to oil with depleted inhibitor into which new inhibitor has been added.

The compounds left behind by depleted inhibitor must be removed from the oil before adding the new inhibitor. A small number of oil recirculations through heat, vacuum and fullers earth will clean up the depleted inhibitor as well as take out any small amount of oxidation decay by-products that may have been absorbed by the insulation.

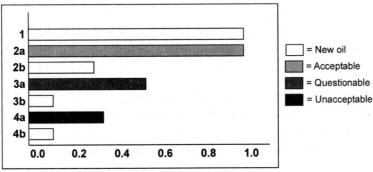

1- New Oil
2a,2b - Oil when inhibitor is just depleted
3a,3b - Oil after oxidation by-products begin to form
4a,4b - Oil after significant oxidation by-product formation
2a,3a,4a - Reclaim and Reinhibit
2b,3b, 4b - Just add more inhibitor

Figure 7.5 - Life expectancy of transformer oil. Bar 1 represents the Life (measured as time of sludge-free operation) of New Oil, given a value of 1.0. If you reclaim and reinhibit after the oxidation inhibitor is depleted, but before acids and sludges start to build up (a more cost effective solution), the oil will have the same life expectancy as that of new oil - 2a. If you don't reclaim, but merely dump inhibitor into the oil, the life of the oil will be dramatically decreased - 2b.

Adequate Inhibitor Content

Retest annually to monitor inhibitor content. As soon as the inhibitor content declines from 0.3% to 0.10%, it's time for the ReInhibit service.

It is very important to ReInhibit your transformers as soon as the inhibitor is depleted and not wait until oil oxidation by-products manifest their presence in the oil via increased acidity or decreased IFT. Why is this so?

Maintaining a proper level of inhibitor in the oil prevents oil oxidation by-products from forming in your transformer. As soon as the inhibitor is depleted, oxidation by-products begin forming. However, they will not manifest themselves in the oil right away because your transformer's solid insulation absorbs them first. Just like the paper filter element in a car's oil filter, the solid, cellulosic insulation in your transformer will actually clean these oxidation by-products out of the transformer oil, as illustrated by Figure 7.5. So, for a period of time you will see no change in the oil's acid number. Oil oxidation by-products damage solid insulation, so this is a condition that you want to avoid. To avoid this condition, ReInhibit promptly when your inhibitor is depleted!

The following graph (Figure 7.6) from a laboratory experiment represents what happens in a transformer after the oxidation inhibitor is depleted. For the experiment we monitored the inhibitor content, acid number and IFT of mineral oil contained in a vessel with a paper-wrapped copper conductor. (Note: The hours shown for the experiment do not relate directly to the timeframe of the same process as it occurs in a transformer.)

A. Inhibitor depleting.

B. Inhibitor depleted (at the 48 hour point). Note that in a vessel containing oil and copper the acid number at this point would immediately begin to climb and the IFT immediately begin to drop. However, the insulating paper at first absorbs the oil oxidation by products.

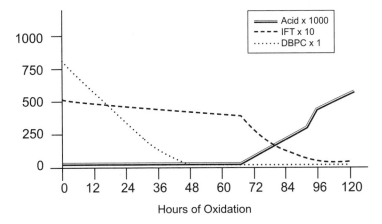

Figure 7.6 - Life of new oil as it relates to inhibitor depletion.

C. The insulating paper has now absorbed enough acids and sludges that some begin to appear in the oil, increasing acidity and decreasing IFT.

D. Continuing oil oxidation. Sludges and acids continue to be absorbed by the cellulose insulation.

Starting from Nowhere

How to Begin a Maintenance Program for a System Suffering from Total Neglect

Many readers come to this work already having a basic understanding of the concepts presented. To some, however, this will all be pretty much new information. The people from whom you received your position of responsibility for transformer maintenance never knew and never taught you about the value of a complete testing program and the value of cleaning and keeping clean the oil in an operating transformer. Consequently, your system contains everything from 30- to 40-year-old transformers with terribly oxidized oil to new units not yet oxidized. Now that you understand the value of extending transformer life through oil regeneration, where do you start?

The first thing to realize is that you cannot make up in one year for a lifetime of neglect. The damage done to your transformers' insulation cannot be reversed even though it can be stopped. And, even though it can be stopped, it is highly unlikely that your organization will find enough money in the budget to clean up all of your transformers this year.

When you realize that you are digging yourself into a hole, stop digging. The simplest, least expensive way to extend transformer life is to make sure that any new transformers you buy have oxidation inhibitor in the oil and are nitrogen blanketed (if possible). As you begin to address your existing transformers, make sure that your organization no longer buys transformers with uninhibited oil. This will at least give you some breathing room with the new units installed in your system. Test these units on an annual basis. When the inhibitor is depleted, reinhibit them. If you follow this plan, you will achieve maximum reliable life from your new transformers from now on.

The next thing to realize is that some of your operating transformers are too far gone to save. Oxidation damage done to oil can almost always be reversed. However, damage done to solid insulation can be slowed or stopped, but not reversed. A competent laboratory can estimate by the oil test results (giving special consideration to the furan test) how much reliable life remains in a given transformer. If a unit is very near the end of its reliable operating life you may opt to replace the unit, and not spend money processing the oil. Here oil is not the problem. The problem is the deteriorated condition of the solid insulation.

Very likely some of the transformers in a system that has been neglected will have reached the end of their reliable operating life – even though they may not have failed yet. You may want to consider beginning the process of replacement for units that testing indicates have reached this point. Even though they haven't failed yet you can't rely on them to withstand the rigors of normal operation. Most transformer owners would rather replace a unit during a scheduled outage than during an unscheduled outage caused by equipment failure. Unless an organization has spare transformers, available work crews and a redundant system that can quickly switch load when equipment fails, replacing a transformer after a failure costs more money than planning ahead. The bills really add up for short notice, overtime work by contractors, rush delivery on equipment and so on – not to mention delivery time (6 to 12 months or more for some units) or lost production if the unit powers some kind of manufacturing operation. The money you save by putting off replacement you lose (and then some) from unplanned outages.

The rest of the system's transformers will fall somewhere in between new purchases with inhibited oil and units that need to be replaced. In order to prioritize servicing these units, the transformer owner will need to consider the following:

Newer Transformers are Less Capable of Surviving Neglect

As we mentioned above, each decade's new transformers are built with less insulation per kVA. That insulation, therefore, must deal with much higher stresses than older transformers. Neglecting newer transformers results in more immediate and greater loss of life than neglecting older transformers.

Which Transformers Would Cost the Most to Replace?

Generally the larger the transformer the more expensive it is to replace. However, other cost factors should not be overlooked. Transformer location can add considerably to the cost of replacement. Whether a basement vault or a roof top substation, some plants seem to have been built around transformers in a way that makes them amazingly expensive to replace. Specialty transformers will cost more per kVA to replace than standard substation units.

Delivery Time

Some transformers you can purchase out of inventory from a manufacturer, broker or even your local utility. You could probably order a 1000 kVA substation transformer with 4160V primary to 480V secondary and take delivery of it later this week if not even later today.

Some transformers have a very long lead time for delivery. If a furnace transformer fails you couldn't even place the order in one day. Delivery of a new unit would occur sometime later in the year if not even the next year. Other specialty transformers have different length delivery times that the transformer owner needs to be aware of. This may directly impact your maintenance decisions.

Significance to Your Operation

A transformer that feeds the parking lot lights of a plant that is open only during the daytime probably is not very significant to your operation. In a utility system that automatically switches the load around failed equipment, a 10 MVA unit somewhere in the distribution system might not be viewed as being a very significant transformer in the overall system.

In any kind of process operation, though – chemical, plastics, metals – the transformers that power the process are definitely significant to the operation. Not only do most plants have delivery schedules to keep their own customers happy but the expense of the operation shutting down in mid-cycle and the difficulty of cleaning up ruined product and starting over definitely can be intimidating. While some units in a distribution system may not be significant to a utility, their generator step-up transformers definitely are significant.

Taking all these factors into consideration, the transformer owner must then exercise his best judgement as to what priorities best serve his organization. A good maintenance contractor may be able to help shape the priorities.

Reviewing the Problem of Sludge

Transformer oil deterioration and sludge formation are two of the major concerns confronting the transformer owner/operator with regards to the insulating oil. Therefore, the key to preventive maintenance is annual oil testing. The primary purpose of annual oil testing is to pinpoint the condition of the transformer's insulation system so that it will operate in the sludge-free range.

The normal aging of transformer oil creates sludge. Sludge appears earlier in life if the transformer is heavily loaded, running hot, abused or neglected. The presence of moisture and oxygen in the oil will accelerate the formation of sludge.

Studies have shown that an increase in the neutralization (acid) number (ASTM D 974-97) should normally be followed with a drop in the interfacial tension (IFT) (ASTM D 971-99) and a darkening of the color of the oil (ASTM D 1524-94). Knowing the acid number and the IFT can give some insight into how much usable life may be left in the oil, the probability of sludge accumulation and what action needs to be taken to prevent or remove sludge deposits.

Definitions

Now that some of the problems that affect transformer oil have been established (moisture, oxygen, sludge, heat), some definitions may be helpful.

1. Reconditioning

    Physical or mechanical methods that remove moisture and/or solid material from transformer oil, including filter presses, vacuum systems and centrifuges.

2. Reclaiming

   Methods that effect a chemical change in transformer oil, such as the removal of oxidation by-products through the use of Fuller's Earth or chemicals. This process is referred to as *regeneration* outside of the United States and has been referred to as *rerefining* in the past.

In addition to these two basic terms, the following are terms frequently used in reference to the treatment of transformer oil:

1. Degasifying

   The removal of dissolved air and other gases, usually with the use of high vacuum.

2. Dehydration

   The removal of dissolved moisture in transformer oil and transformer cellulose insulation. This is generally accomplished with elevated heat and high vacuum..

3. Hot Oil Clean®

   The removal of sludge and oxidation by-products from transformer insulation and from the inside of dirty transformers. This process is done by heating naphthenic transformer oil up to its aniline point; the point where the oil will act as a solvent on the sludges it comes in contact with. The passing of the dissolved sludges through an adsorbent filter bed traps the sludge, removing it from the oil, paper and tank.

4. Reinhibiting

   The addition of an antioxidant such as DBPC® to transformer oil. This process is usually done when all oil test results are acceptable but the inhibitor level of Type II (inhibited) oil has dropped to 0.09% by weight.

5. Adsorption

   A process in which one substance attracts and holds another substance to its surface area (like metal filings attracted to a magnet).

6. Absorption

   A process in which one substance is soaked in or up in another (like water in a sponge).

Confronting the Sludge Problem

Introduction

Many transformer experts realize the problem of sludge in operating transformers. Sludge forms when acids attack the iron, copper and varnishes. The by-products of these reactions go into solution and eventually precipitate out of solution as sludge. Sludge adheres to the cellulose insulation, the sidewalls of the tank and lodges in cooling tubes and ventilating ducts. Sludges are adsorbed into the cellulose fibers of the insulation even before it precipitates out onto interior surfaces. Once sludge adsorbs in the cellulose, the degree of polymerization and tensile strength starts to decline, the formation of furanic compounds accelerates and the cellulose insulation begins to lose its plastic quality. Since this sludge problem is one of the prime causes of premature mechanical failure of transformers, something must be done to correct the problem before imminent transformer failure.

Seven Alternative Approaches

The goal of the transformer owner/operator is to keep the transformer operating in the sludge free range. There are seven approaches to keep sludge from appearing in transformers and to deal with sludge that is already in transformers.

- Ignore the situation
- Recondition the oil
- Retrofill the oil
- Reclaim the oil
- Hot oil clean the transformer
- Reinhibit the transformer oil
- Fluidex technology

Ignore the Situation

Most transformer owner/operators have some working knowledge of the problems that sludge can cause in transformers. The ignorance of methods available to deal with sludge is no longer a excuse. This attitude is mostly historical.

Recondition the Oil

There are many reconditioning methods available. One of the first known methods dealt with the first problem in oil maintenance, moisture. A mechanical press known commonly as a filter press has been used for oil problems for quite some time. Porous paper filters absorb free water and some dissolved water, while trapping particulate matter. They do not trap dissolved oxidation by-products in the oil (see Figure 7.7).

They do not solve the problem of sludge in the transformer. Notice also that the cellulose insulation in the transformer absorbs material out of the oil just as the paper filter element in a filter press. Keep in mind that a transformer will have many times more paper in it that the filter press does!

It is very difficult to gauge when the filters have absorbed all of the moisture that they are capable of holding. When these filters get too wet, they can disintegrate and release cellulose fibers into the oil. These fibers should not be in oil that will be used for in-service equipment.

Cartridge filters work along the same lines as a filter press and have the same affect (or lack thereof) on the transformer. Cellu-

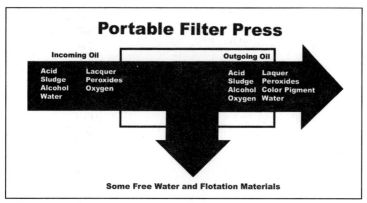

Figure 7.7 - Results of Filter Pressing the Oil

lose fiber filters are used but a bit more engineering goes into the making of the filter. The cartridge filters are generally used in an enclosed tank of some kind, so the environment (humidity, dust) will not affect the filters.

A centrifuge is another reconditioning method used for separating gross amounts of water and carbon. It is used for cleaning bulk quantities of oil and tends to aerate the oil. Centrifuging leaves about one percent (10,000 ppm) water in the oil, so it is not a very effective reconditioning method. Additional processing will be required, so why bother using it in the first place? Again, nothing is being done about sludge utilizing this method.

Coalescing filters act somewhat like a centrifuge but they have no moving parts. Oil with dispersed water enters the coalescing fiberglass element, which traps small water droplets. The increasing pressure differential across the filter medium forces the droplets of water together. They eventually get large enough to pool at the bottom of the filter, where the water can be drained. Dry oil passes through a separator screen (see Figure 7.8).

This filter removes all free and dispersed air. Because the effluent oil will have a low water content, coalescers are used as a prefilter for vacuum or fuller's earth treatment. A major application for these filters is to remove moisture from jet fuel at airports.

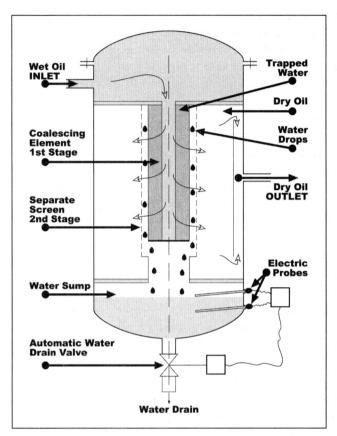

Figure 7.8 - Coalescing filter/water separator

Vacuum dehydration is a highly efficient means of reducing the gas and moisture content in transformer oil as well as removing the volatile acids present in the oil. The chief function of this method is water removal. The oil is not treated chemically and the sludge problem in a transformer still has not been addressed. Here heat will work with vacuum to remove water more efficiently. Using a filter press, heat will dissolve the water and keep it from being removed by the filter element.

Retrofilling the Oil

The process of draining down deteriorated oil, flushing the core and coils and refilling with new oil has been around for a long time. Flushing down a dirty transformer reaches only 10% of the interior surface. Soon after the new oil is added, the contamination level will return to close to its original state. The new oil gets this bad because 10% of the volume of oil in a transformer becomes trapped in the cellulose insulation. Putting a small amount of oxidized oil in with new oil can ruin a large quantity of new oil.

Retrofilling oil does not remove deposited sludge on the core and coils and is not a solution to the sludge problem. Disposal of the removed oil presents its own set of problems.

Reclaim the Oil

The process of reclaiming is simply the passing of badly deteriorated transformer oil through an adsorbent bed. This process removes most oil oxidation decay products, polishes the oil and restores oil to like-new quality.

Most of the contaminants in oil, including water, are polar in nature and are therefore easily adsorbed. Several types of adsorbents are available. The most widely used is fuller's earth. Activated alumina, trisodium phosphate and activated carbon can also be used.

Fuller's earth refers to a class of naturally occurring adsorbent clays rather than to a specific mineral species. The main constituent in this class is attapulgite clay, mined principally in southern Georgia and northern Florida. It is most effective for its use in transformer oil reclaiming because of its abilities to adsorb polar compounds and to lighten the color of service-aged oil.

What makes attapulgite so unique is its crystalline structure. As mined, the clay is a hydrated magnesium and aluminum silicate. During processing, the clay is crushed, heat-activated, ground, screened and bagged. The temperature of the heat activation or drying stage determines the degree of internal porosity. This porosity contributes to the clays high surface area and adsorptive capacity. Burning or calcifying at temperatures between 800 - 1000 °F activates the clay. This treatment develops porosity in clay as high as 13 acres per pound of clay (125 m²/gram).

In order to maximize speed of adsorption, the smallest granular size should be used. In the 30-60 size, there are 30 holes per inch in a screen and every grain of earth will pass through the screen. At the upper limit of 60, none of the grains will pass through the screen in the 20-minute screen analysis. On the other hand, the larger the grain sizes the slower the adsorptive capacity.

Activated fuller's earth can adsorb large quantities of acid. (Figure 7.9) The amount of acid removed depends on many factors since adsorption is a dynamic equilibrium process. Temperature, flow rates, viscosity of oil, residence time and initial level of acid all affect the rate and capacity for adsorption.

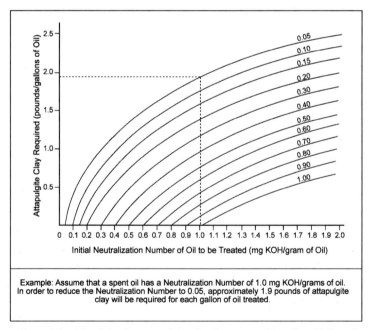

Figure 7.9 - Adsorbtion by attapulgite clay (courtesy of Engelhard Mineral and Chemical Corporation)

Reclaiming using fuller's earth is accomplished by contact at an elevated temperature with powdered clay or through percolation through granular clay, using pressure or gravity to force the oil through the clay (Figures 7.10 and 7.11).

Heat thins the oil's viscosity. There are methods for using both of these procedures. In-depth information about contact and percolation methods can be found in IEEE Std 637-1985, *IEEE Guide for the Reclamation of Insulating Oil and Criteria* for Its Use and IEC Std 60422-1989, *Supervision and Maintenance Guide for Mineral Insulating Oil in Electrical Equipment.*

Hot Oil Clean® the Transformer

Simple reclaiming of transformer oil is not the complete solution to oil maintenance. While polar contaminants are adsorbed and removed from the oil, reclaiming doesn't address the issues of moisture and dissolved gas. Sludge does not exist in just the oil but must also be removed from the cellulose insulation and transformer tank.

What is needed to clean all of the insulation of sludge is a solvent. Fortunately, hot transformer oil is a solvent for its own decay products. The degree of heat necessary is indicated by ASTM D 611-82 (1998), *Standard Test Methods for Aniline Point and Mixed Aniline Point of Petroleum Products and Hydrocarbon Solvents.* This method measures the temperature that the oil dissolves aniline, an aromatic substance. The degree of mutual solubility increases with temperature.

ASTM D 3487-00, *Standard Specification for Mineral Insulating Oil Used in Electrical Apparatus,* lists the aniline point at 63-84 °C. When heated to this temperature, transformer oil becomes an effective solvent for its own decay products. Hot Oil Cleaning combines the features of reclaiming plus the solvency power of the hot oil to attack the sludges within the transformer itself. Hot Oil Cleaning consists of three steps; heat, adsorption and vacuum. Thus, reclaiming cleans the oil, heat/solvency eventually removes

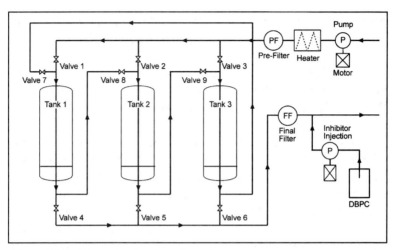

Figure 7.10 - Schematic of modern three-clay tower system.

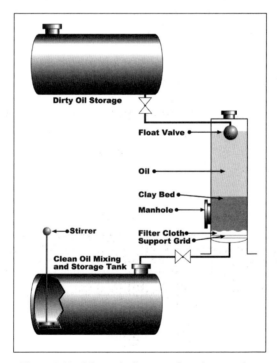

Figure 7.11 - Schematic diagram of gravity-percolation reclaiming apparatus.

oxidation by-products from the tank and the cellulose insulation and vacuum removes moisture and gas. (Figure 7.12)

This allows transformers to operate in the sludge free range, prevents premature failure due to loss of mechanical strength of the insulation and conserves two resources, oil and transformers.

While transformer oil can be reclaimed in just one pass of the oil through the equipment, it takes more passes to sufficiently clean the transformer and the cellulose insulation. In most cases, the work is done safely and more efficiently while the transformer being processed is energized. The cleaning is more efficient because the 100/120 cycle coursing through an energized transformer causes the laminations to vibrate. The vibration helps loosen sludge from deep inside of the system. The heat produced in an energized transformer will also aid in getting the oil temperature to the aniline point in a more timely manner, causing the heater to work more efficiently (Figures 7.13 and 7.14).

Reinhibiting the Transformer Oil

Reinhibiting transformer oil is a process that renews the concentration of oxidation inhibitor (DBPC or DBP) to ASTM Type II level of 0.30% by weight. The same equipment used to Hot Oil Clean® transformers is used to reinhibit the oil. Reinhibiting prevents the acceleration of oxidation by-products forming in ser-

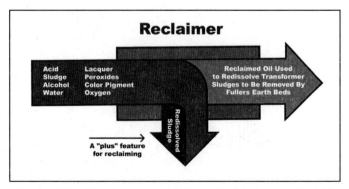

Figure 7.12 - Showing contrast between Reclaiming and Portable Filter Press for transformer oil treatment.

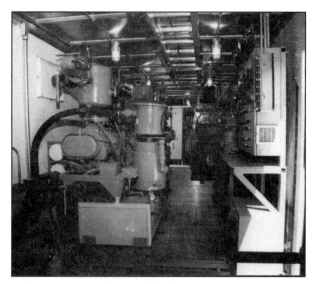

Figure 7.13 - Inside View of Reclaiming Rig

Figure 7.14 - Fullers' Earth Tanks

viced aged transformer oil, thereby providing life extension for the high-grade naphthenic-based oil already in service. Due to these factors, the mechanical strength of the cellulose insulation is also maintained.

Reinhibit is performed on transformers that have acceptable oil test results in all areas except inhibitor level. By adding inhibitor at this time, the formation of dangerous sludge is halted. Any oxidation by-products that may have formed are eliminated when passed through the fuller's earth beds and any residual moisture and gas is removed with the vacuum.

Fluidex Technology

If there is a drawback to Hot Oil Cleaning, it is with the fuller's earth. Fuller's earth can only adsorb so much polar contamination before the beds of fuller's earth must be changed out. If they are not changed out, oil with a high neutralization number will be introduced back into the transformer. Once the earth tanks are renewed, Hot Oil Cleaning will resume. You are now faced with a disposal problem with the spent fuller's earth, which contains residual transformer oil.

Oil soaked fuller's earth cannot be dumped in sanitary landfills. The potential of groundwater contamination with these petroleum hydrocarbons is not a problem that industry wants to deal with. Therefore, spent fuller's earth must be drummed up and sent to a proper disposal facility, which will run a waste profile on the oil soaked fuller's earth and charge a lot of money for the disposal of the product.

Fluidex technology offers many advantages over traditional Hot Oil Cleaning systems. With Fluidex systems, fuller's earth beds can be used 200-300 times before they need to be replaced. This reusability factor eliminates the need and the costs associated with disposal, extra handling of material, environmental issues of dumping oil-soaked fuller's earth, and the special equipment that may be necessary for dumping fuller's earth.

Fluidex technology reduces oil losses associated with the dumping of spent fuller's earth by well over 90%. Fluidex technology also may reduce oil-processing time by 20-70%. The fuller's earth used in traditional Hot Oil Cleaning has much higher moisture content than that used by Fluidex. This being the case, less oil processing is required to properly dehydrate the processed oil.

Fluidex is capable of Hot Oil Cleaning the transformer and reinhibiting the transformer oil without the hassles of fuller's earth disposal, without potential ground water contamination from residual transformer oil and without depleting supplies of fuller's earth. With these things in mind, oil savings and cost savings due to fullers earth purchase and disposal, fluidex technology can prove to be a more cost effective alternative method in oil reclamation (Figure 7.15).

Figure 7.15 - Fluidex Plant

Factors to Consider

Regardless of the method used for treating in-service transformer oil, these four critical factors should be incorporated into the chosen equipment:

1. Dehydrate the oil before it contacts the adsorbent in order to prevent water from wetting the filter material. Water will breakdown the crystalline structure, making it necessary to discard that batch of filter material.

2. Dehydrate the oil coming out of the filter material to prevent moisture from being present in the treated oil. For

example, fuller's earth can hold 10-13% moisture, which can be passed on to the treated oil. This is particularly true when recirculating oil in a transformer.

3. Filter oil through a 0.5 micron filter prior to returning it to the transformer. This will remove any remaining filter material, lint, fine ground iron, or other contaminant that passed through the processing equipment.

4. Add an oxidation inhibitor to reclaimed oil in concentrations up to 0.30% by weight. Processing oil through filter material at 70 °C or higher and through a tight vacuum can remove oxidation inhibitor left in the oil prior to servicing. Extremely oxidized oil has also had its inhibitor consumed in the deteriorating process.

Conditions that have proven satisfactory for most inhibited mineral oil processing are found in Table 7.2. These are conditions that will not remove the inhibitor as the result of processing.

Energized Hot Oil Cleaning vs. De-Energized Hot Oil Cleaning

Most transformers can be desludged through hot oil cleaning while the transformer is energized. There are some exceptions. First and foremost is the safety issue. If the reclaiming operator feels that the transformer cannot be serviced safely, it should not be done unless the transformer can be shutdown. Cabinet-type transformers where the bottom valve is very close to the energized leads fit this category quite often, as do transformer banks on raised platforms and pole-top transformers (transformers with no valves) with primary voltage over 7200V.

Other considerations that would prevent a transformer from being hot oil cleaned while the transformer is energized are certain oil characteristics. A percent moisture by dry weight of over 6% (a temperature where flashover could occur at 55 °C) and a dissolved gas level that is potentially dangerous, will prevent a transformer from being hot oil cleaned on an energized basis. These ranges will be identified through various oil tests.

ASTM Recommended Processing Conditions of Inhibited Oil[1]

Temperature °C[2]	Minimum Pressure	
	Pascals	Torr (approximate)
40	5	0.04
50	10	0.075
60	20	0.15
70	40	0.3
80	100	0.75
90	400	3.0
100	1000	7.5

[1]ASTM D-3487-00, Note 1
[2]If higher temperatures are used, test the oil for inhibitor content and add as necessary.

Table 7.2

Regardless of the choice of method, when energized hot oil cleaning will be performed, only companies who specialize in this kind of work should do it.

When Will a Hot Oil Cleaned Transformer Require Additional Treatment?

While the initial hot oil cleaning procedures (measured in hours) will remove large quantities of decay products, only a time-exposure of months will effect the complete removal of residual sludges. Hot oil cleaning that takes 30 hours cannot fully correct a problem that developed over a 30-year period. The longer that hot oil cleaning has been put off, the more sludge accumulation dictates that more time is needed for the Hot Oil Clean® process. As a guideline, and to stay ahead of the sludge problem, monitor the transformer oil annually after the initial hot oil cleaning. Re-service the transformer when the acid number and interfacial tension results fall into the questionable parameters. This will provide many more years of reliable service.

Transformer oil test data on transformers that have been completely desludged show that hot oil cleaning procedures effectively solve the problem of removing extensive sludge build-up in transformers. This data has been accumulated for over 30 years and involves oil test results from tens of thousands of transformers.

Other Constraints of Energized Hot Oil Cleaning

As briefly mentioned in previous sections, most of this kind of work can safely be done while the transformer is energized. However, there are some constraints that would prevent hot oil cleaning from occurring while the transformer is energized. Some of these things are:

Retrofilling or Inadequate Reclaiming

Figures 7.16 - 7.18 illustrate what happens when a transformer with sludge deposits receives inadequate servicing. Inadequate service includes both replacing old oil with clean oil (retrofilling) as well as too few passes of oil reclamation (regeneration).

All three figures show sludge deposits (in black) in typical locations throughout the transformer. Keep in mind, though, that the concept illustrated on this page includes more conditions than just transformers with visible sludge. Any transformer with oil oxidation by-products in the oil also has oil oxidation by-products ("sludge") deposited in the cellulose insulation – whether visible or not.

Oil Test Analysis

Energized hot oil cleaning should not be attempted if liquid screen test results reveal cloudy insulating fluid, which may indicate excessive moisture. Energized work should not be done if the visual examination shows sediment-containing metallic content or bits of cellulose insulation.

Figure 7.16 - Here we see an illustration of a transformer with heavily oxidized oil that has deposited sludge on the transformer's internal surfaces – at the top of the core and coil assembly as sludges precipitated out of the oil and on the cooler surfaces at the bottom of the coils and along the radiator tubes. The wrong approach to correcting this problem would be to replace the oil.

Figure 7.17 - This graphic illustrates the results of this approach. As soon as you replace the oxidized oil with clean oil, the transformer is filled with clean oil. No surprises here. However, the sludge deposits remain untouched! Too few passes of reclaiming or reclaiming at temperatures below the aniline point produce the same results.

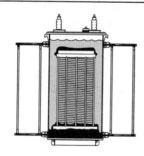

Figure 7.18 - This graphic illustrates the next stage after retrofilling or improper reclaiming. After replacing the oil or after inadequate reclaiming, the sludge deposits redissolve in the clean oil. In a short period of time you will have contaminated oil again along with the untouched sludge deposits.

Hot oil cleaning must be done with the transformer off line if Karl Fischer moisture tests indicate a percent moisture by dry weight of over six percent for transformers over 500 kVA or over 50 ppm for transformers less than 500 kVA.

If dissolved gas analysis indicates high concentrations of combustible gases, hot oil cleaning must not be done while the transformer is energized. Repairs and degassing need to be done before any energized servicing. Acetylene indicates arcing. Any transformer with arcing problems should not be treated at all until the transformer has been taken off line, repaired and the oil degassed.

A high furan content is indicative of cellulose insulation degradation. If the furan content in the insulating fluid is too high, it may be recommended to de-energize the transformer before servicing.

Observations

There are some observable things that need to be considered for energized hot oil cleaning of transformers. A terminal board on the tap changer assembly that has visible sludge should not be serviced while the transformer is on line.

A pole-top transformer over 7200 primary volts needs to be taken off line while the oil is being processed. It is necessary under this condition to introduce hoses through the inspection plate at the top of the transformer. This plate is usually located between the high voltage bushings. Besides this obvious danger, other problems could be insufficient clearance for the hose between the windings and the transformer case and the possibility of free water at the bottom of the transformer with no way of detecting it.

Transformer Age

Transformers that have been in service for fifty years have been successfully hot oil cleaned while energized. However, most transformers don't live to be that old. When considering energized servicing of an older (+50 years) transformer, determining if there is sufficient life left in it to justify the cost would be in order. Degree of polymerization and Furan content will help make this decision the right one.

Recognizing that the Sludge Problem is Fixed

It will be easy to recognize that the sludge problem has been solved with hot oil cleaning. The transformers will operate cooler, sometimes in excess of 10 °C cooler. Transformers under normal operating conditions will require only annual oil testing and occasional reinhibiting of the transformer oil. The transformer will enjoy extended life, free from accelerated degradation of the cellulose insulation.

The Reliability of Energized Transformer Hot Oil Cleaning

Almost 40 years of field experience has proven the prediction by Frank Doble in 1952, that energized servicing of transformer oil would be safely accomplished if the proper equipment is used. Since 1965, S. D. Myers, Inc. has successfully serviced tens of thousands of transformers on an energized basis. Of course, specific safety procedures are followed.

Safety Precautions

Safety precautions begin with oil testing and analysis to determine if there are any unusual conditions in the transformer that might make it inadvisable to Hot Oil Clean® while the transformer is energized.

Basic safety precautions for personnel protection must include the following:

- Observe OSHA rules for training to enter substations.

- Maintain a safe and clear distance from energized conductors.

- Assure adequate grounding of the processing equipment to the substation ground.

- Use wire-braid-reinforced high-pressure hose for all connections between the transformer and the processing equipment.

- Ensure that all hoses and connections are bonded to prevent static discharge.

- Provide the circuits serving booster pumps with ground-fault circuit interrupters (GFCIs) to protect personnel from shocks.

- Use insulated gloves, blankets, sleeves and other personal protective equipment where necessary.

Energized Processing Innovations

The following safety features should be built into the processing equipment system and their operation monitored by operating personnel:

- Liquid-level detectors and controls to ensure that current-carrying internal transformer components remain under oil. These controls should be designed to maintain a constant oil level in both the transformer and the process equipment.

- Automatic heat regulation to maintain the oil at the aniline point and ensure that the temperature will not rise high enough to damage the insulation system.

- Alarm devices to signal when temperature, pressure and oil levels deviate from prescribed limits.

- Filters at the discharges of all pumps to catch metal cuttings that may be released from pump impellers and gears.

- Design that ensures that vacuum is drawn only in the process equipment and not in the transformer. (Never pull a vacuum on an energized transformer.)

- Instrumentation for monitoring flow rates to provide an evaluation of the fuller's earth filter efficiency. A log of flow rates should be maintained throughout the Hot Oil Clean® process. Oil must be sampled and tested to determine its condition. Tests should be conducted using the latest ASTM test method.

Energized Reclaiming – Objections Answered

Some transformer authorities assert that there is no advantage to energized reclaiming, especially when you have zero risk with de-energized work (if you can arrange to have the transformer shut down). Three risks are commonly cited as reasons for not servicing an energized transformer:

1. Negative Transformer Tank Pressure (Vacuum)

   One argument for de-energized work is that vacuum could be drawn on the transformer, causing collapse of the tank or cooling tubes. This is an invalid concern if the process equipment draws vacuum only in the process equipment itself. The transformer headspace must be vented to the atmosphere throughout the reclaiming, thus eliminating this objection.

2. Air Bubbles

   Another argument for de-energized work is the development of air bubbles in the windings or in the reclaiming process causing air lock in the pumps and flashover in the transformer. However, a properly designed system uses double protection against this possibility in the form of an air trap and vacuum degasification coupled with a controlled flow rate. The oil returned to the transformer has a tendency to absorb gas, not produce it.

3. Stirred Up Contamination

   The last objection to be raised is that stirred up particulate matter in the transformer will negatively affect the dielectric breakdown strength of the transformer oil. This concern is unjustified because of three fundamental features of the Hot Oil Clean® process:

   - There is no forced circulation in the process of servicing an energized transformer.

   - Oil is removed from the bottom of the transformer tank and returned at the top. The reclaiming process is done gradually over a period of time. The hot processed oil "floats" on the cooler oil in the transformer as the viscosity of the hot oil is less than the cooler oil.

   - There is a clear cut interface, the processing equipment eliminates the recirculation of contaminants in the transformer tank by trapping them in filters or a fuller's earth bed.

Once contaminants in suspension have been drawn off at the bottom of the transformer tank they are gone forever. Only clean, dehydrated, degassed, particulate free oil is returned to the transformer. Hot oil cleaning is a process of gradual dissolution of sludges over a period of time, with contaminants going into solution as they are liberated from the inner surfaces of the transformer. Therefore, there is absolutely no agitation of particulate matter during the processing of energized transformers. There is only a continuous polishing of the oil in the reclaimer and the dissolving of soluble sludges – oil decay products.

Visual Proof in Helping to Make a Valued Decision

Up until 1985 we were quite satisfied with the oil test parameters we were using. In fact the IEEE Std 637 published that year, essentially put a stamp of approval on them. There were still problems to solve. Transformers still had sludge problems. The old parameters were based on old research and data from 1957. What were the changes in transformer design since the parameters were established? Thermal upgrading of the transformer with a designed hot spot temperature change to 110 °C. Higher electrical stresses (kV/cm) with newer designs. Computer aided design, reduced BIL levels. Then we learned about the (bad) characteristics of Kraft paper. That it has absorption and adsorption properties, it is an acid former, it has plasticity under pressure and it's a catalyst. All of these things needed to be taken into consideration when establishing oil-testing parameters.

We then ran some experiments. Kraft paper was taken and aged at various acid levels and compared under a microscope. This was done a magnification levels from 100x to 1000x. Here is one series at 750x, what do you see?

I think you can see a clear progression toward disaster! The view depicting the new oil and paper and the oil and paper with a 0.05 acid number looks just fine. In the view of the oil and paper with a 0.10 acid number we start seeing debris floating around. What is it? Sludge! The next view, 0.15 acid number, we see much more sludge with paper starting to delaminate. And the final view, 0.30 acid number, so much sludge you can't see the paper! There are some standards out there that say a 0.50 acid number is ok!

New Oil

Acid # 0.05

Acid # 0.10

Acid # 0.15

Acid # 0.30

The job of protecting the paper and the mechanical strength of the transformer starts with proper oil maintenance. Reinhibit® and Hot Oil Clean® your oil-filled transformers when indicated by the Neutralization Number Test (Acid #).

Conclusions

With almost 40 years of experience, we have established an in-depth data base on many thousands of desludged and hot oil cleaned transformers. This database confirms that using the proper equipment and procedures, complete desludging of energized transformers is economical and a safe procedure that prolongs transformer life. It also results in the preservation of quality naphthenic-based transformer oils.

To reiterate, transformer mineral oil itself can be reclaimed in one pass through a fuller's earth bed of proper size and utilization. Additional passes of hot naphthenic oil required by hot oil cleaning procedures are to dissolve decay products in and on cellulose insulation, core and coils.

The number of recirculations through the system required to accomplish the cleaning of a transformer depends on the equipment used, the amount of fuller's earth, the condition of the oil being treated and the age of the transformer being serviced. It is recommended that a minimum of 1500 pounds of fuller's earth is used and oil temperature is at the aniline point as it enters the top filter press valve. This work must be done no later than when the acid number and interfacial tension reaches the questionable range.

Chapter 8

Remanufacturing

Even the best maintained transformer will eventually reach the end of its reliable operating life. At that point, the owner must decide whether to repair or replace it. Economic factors will direct the decision. However, the economics of the decision can be calculated in different ways depending on the aims, values and present economic realities of your organization. Different decision-making processes will favor different decisions.

For instance, some utilities don't even consider rewinding any transformers under 10 MVA or 25 MVA. Some don't consider rewind for any units under 50 MVA. In practice, these transformers are treated like light bulbs. When they burn out, replace them. Often this policy is not supported by any kind of strong, financial analysis. It may only spring from a perception that it is easier to buy a new transformer than it is to repair an existing one. A new, competitive environment is causing many of these utilities to rethink these policies.

A rewound, remanufactured transformer should provide an equally long reliable operating life as a new transformer. In fact, since newer transformers continue to be smaller, using less and less insulation with greater stressing of materials, a rewound, older transformer might have a longer operating life than a new transformer.

Additionally, rewound transformers usually cost less than new transformers and have shorter delivery times. This only stands to reason since much of the transformer is being reused, not built from scratch – the tank and the core, for example. The cost savings of rewinding vs. buying new is often in the 25-40% range. The savings on delivery times is often more significant than the cost savings. Some specialty transformers (for instance, furnace transformers) and larger transformers may have delivery times in excess of one year for new units. This can give the rewind option a tremendous advantage, particularly if delivery time is important to your operation.

These days an additional consideration for choosing remanufacturing is the inability to have new shell form and very large core form transformers built in the United States. With most manufacturing facilities in other countries, it becomes more difficult during the purchase, design, design review, construction, and testing process. It will also add a significant amount to the total cost of the transformer due to shipping costs.

Some options can further reduce the cost of a rewind but jeopardize the integrity and longevity of the transformer. One such option is the use of aluminum conductor. It may cost less up front, but you get what you pay for. With aluminum-wound transformers you get less of a transformer. See p.4 for further discussion on aluminum vs. copper windings.

The same holds true for layer-wound transformers. Again the up-front cost is less but you get a transformer that cannot stand up to the same stresses that a disk-wound unit can. See p.5 for further discussion on layer- vs. disk-wound construction.

In some installations, rewind provides the advantage of keeping a fit between the existing transformer and equipment surrounding its installation. For example, many transformers sit between switchgear lineups. New transformers won't fit the throat connections. Making a new transformer fit the old switchgear lineups could be time-consuming and costly. Rewinding the existing transformer or an identical sister unit avoids this extra time and expense.

An older transformer that reaches its end of life very likely was designed before the advent of computer-aided design (CAD). Many older transformers can be uprated from their existing kVA class using better design and stronger materials – in particular, thermally upgraded solid insulation (see p.21).

Rewind often does not provide an attractive alternative to buying new for smaller distribution transformers (generally pole- and pad-mount transformers < 500 kVA). The mass production manufacturing process for these units is so efficient that it is hard to beat in the repair shop. A transformer owner may find an advantage in price and/or delivery for small specialty transformers, though. As well, if immediate delivery is your highest priority and the unit you are looking for is a common size with common voltages, electric supply companies may have a new unit in stock and available for immediate delivery.

Overall, rewind is a legitimate option that should be considered seriously when a transformer reaches the end of its reliable operating life.

Reusing Transformer Core Steel

The cores of power transformers are constructed of thin gauge silicon steel laminations (0.009" to 0.014" thick) to limit the flow of eddy currents within the core assembly. The laminations are treated to develop a chemical coating (carlite) that insulates the laminations from each other. This film provides a high interlamination resistance that protects against excessive energy loss in the core.

The carlite coating is mechanically durable and is not readily damaged during handling. Thermally, the carlite coating is not affected at temperatures up to 1600 °F (870 °C). Operating temperatures of transformer cores rarely exceed 400 °F (204 °C).

When a transformer fails during the warranty period and is repaired or when a transformer is remanufactured, the core laminations are dismantled to access the windings and are stored for reuse. Most transformer failures do not experience core damage. Minor lamination damage is repaired and the laminations are reused. Care is taken to inspect the laminations during core dismantlement. The laminations are not bent or dropped during handling. Damaged laminations are identified and packaged for repair or replacement.

The remainder of the core laminations are wiped clean of debris, placed on pallets, and stored in an environmentally controlled factory area.

During core rebuilding the width of core lamination gaps are quality controlled not to exceed 3/16 inches. In some cases the core loss is lower than original core construction values, but on average the core loss is approximately 2% to 3% higher than the original value.

Transformer industry practice is to reuse core steel laminations. If the integrity of the laminations is in question, a Franklin test can be performed to determine the quality of the surface insulation. An extended core excitation test, while monitoring for combustible gases, will also prove that a transformer core will operate satisfactorily in service.

Failure Options for Large Shell Form Power Transformers

When a large shell form power transformer fails, especially an extra high voltage unit, (345 kV and above), a number of options are available. They include:

- Repair consistent with the original design.
- Remanufacture to state of the art.
- Purchase a new core form/shell form unit and scrap the original unit.

Remanufactures are capable of remanufacturing large shell form transformers to state of the art. This as an optimum solution rather than repairing (copying) the original coil and insulation design. Modern materials and design technology as well as accumulated historic failure information is now available which allows us to correct the shortcomings of original equipment transformers.

Most often they are able to effectively remanufacture a shell form transformer for considerably less than the cost of a new core form or shell form unit. In many instances they can provide lower losses than the original unit's certified test report while still maintaining the required design parameters. With the generous core window spacing present in most original shell units, remanufactures can design an updated insulation system with greater dielectric margin and higher mechanical strength than provided originally. Seeking more economical units, O.E.M.'s continually downsize their new designs to smaller electrical spacing by developing complex insulation systems that remain to be proven in service. We think downsizing has some serious limitations.

To properly evaluate your options, review the following important factors and incorporate them into your final decision.

- In these fast-changing times, many utilities are now focusing on initial outlay of money for transformer replacement rather than estimating a total cost of ownership, including the cost of losses over a 30-year period. Perhaps your evaluation of ownership needs to be updated to a less forward thinking analysis and more to present savings.
- A high confidence level in the vendor's electrical and mechanical transformer design technology to produce a modern custom designed shell form transformer.
- Degreed engineers and experienced drafting personnel located at the facility to analyze failures and prepare detailed engineering reports for education and documentation.
- Knowledgeable personnel available for gathering data and performing other technical observations in the field who can offer advice with substance.
- If purchasing from a foreign supplier is appealing, consider shipping damage, warranty or after warranty support requirements.
- When delivery is extremely important, perhaps a union shop should be avoided if contract negotiations are of concern.

- Inquire about a sufficient number of factory associates trained and experienced in the dismantlement and rebuilding of large shell units.
- The facility should have suitable winding machines and special tooling, which are designed specifically for shell form transformer remanufacturing.
- Modern transformer processing using vapor phase for removing contaminants does a faster, more efficient job than outdated oven drying methods.
- Dielectric, temperature test and the remainder of tests outlined in the current ANSI/IEEE standards at full nameplate rating are preferred to predict design performance.
- Ask about the facility's success rate as to factory and field failures, especially 345 kV and above.
- Inquire about references that may be approached as to their satisfaction of work performed by the facility.

We think a wise decision is to remanufacture most large shell form transformers to state of the art. We justify this recommendation with the economics and proven service life of this style of transformer. Inquire about the advantages of shell form construction and upgrades that correct problems involving design and building inadequacies of various vintages of McGraw Edison, Cooper Power Systems, ABB, Westinghouse, and Allis-Chalmers shell form transformers.

Shell Form Power Transformer Corrected Problems

McGraw Edison, Cooper Power Systems, Westinghouse, ABB and Allis-Chalmers have all manufactured shell form transformers for utility and industrial service. Each of these manufacturers experienced specific in service and failure problems due to their own unique design and construction. Using state of the art design techniques, materials, and manufacturing methods, Ohio Transformer, has successfully remanufactured all of the above vendors' shell form transformers, having identified most original design and building inadequacies.

The problems found and corrected in the above shell form transformers include:

- Burning and deterioration of inner electrostatic shield grounding straps located in high leakage flux fields
- Magnetic shielding inadequacy
- Coil turns not properly supported by washer/spacer assemblies and other insulation enclosures that are subject to fault current movement
- Insulation overheating (burning) of tap and coil finish lead insulation when these items are located in high leakage flux fields
- Core steel laminations contacting "T" beams at ground potential, producing an unintentional core ground resulting in formation of combustible gases
- Coil to coil braze joint overheating caused by improper brazing of conductor groups and insulating methods used
- Inadequate ventilation of core to eliminate overheating of core steel laminations
- Improper location of coil conductor transpositions to control circulating currents in windings
- Incorrect blocking and insulating of coil conductor transpositions resulting in incipient failure
- Coil to coil brazed joint failures caused by poor quality torch brazing methods
- Poor core building tolerance resulting in excessively wide core joint gaps causing high exciting current/losses and joint heating
- Moisture contamination due to bad processing
- Electrostatic shielding (static ring) improperly designed or manufactured
- Static electrification with probable origin being a combination of oil temperature, high velocity oil flow and turbulence caused by high pump volume and cooler sequence of energization
- Units utilizing a 2-inch coil radius at inside corners increasing the effects of top core insulation, inside turn, and core lamination heating due to elevated leakage flux densities

- Sagging bottom "T" beam that causes "T" beam magnetic shielding to contact bottom tank wall magnetic shielding, thus promoting a potential source of combustible gas formation
- Bottom tank electrostatic shielding breaking loose and drifting up toward bottom of phase package resulting in potential failure of transformer
- Top tank wall magnetic shielding coming in contact with upper "T" beam magnetic shielding in two piece tank construction, thus promoting a potential source of combustible gas formation
- Excessive burrs on edges of core steel laminations causing induced eddy currents to form, increasing the potential for oil sludging and/or combustible gas formation
- Mechanical Failure of internal tank structural members, during external transformer faults, perpetrating coil movement and electrical failure

Chapter 9

Disposal

The first consideration for disposal of a transformer is its PCB (polychlorinated biphenyl) level. The benefit of scrap value recovery of the valuable metals contained in a transformer – copper, aluminum, brass, steel - must be weighed against the potential environmental liability from sloppy or improper disposal and recovery methods.

PCB was used as an insulating fluid in transformers for applications where flammable insulating fluids were not acceptable – such as locations in, on, or near buildings. Ordinary transformer insulating mineral oil is fairly close to diesel fuel in its ability to sustain a fire once ignited. A sustained arc in a failing transformer can very easily ignite the mineral oil insulating fluid. PCB provided a less flammable insulating liquid. It was widely used for this purpose until environmental regulations around the world banned its manufacture and regulated its use and disposal.

Mineral oil-filled transformers were never designed or intended to contain PCB. However, over the years PCB has found its way into many mineral oil -illed transformers. Both PCB and non-PCB transformers may have been repaired in the same shops or had their fluids processed by the same equipment. Once some PCB found its way into a few mineral oil transformers, oil handling and processing equipment spread it around to many others.

For our purposes we will consider three PCB levels: transformers containing pure PCB, mineral oil transformers containing some PCB, and mineral oil transformers containing no PCB.

United Nations Environment Program
Stockholm Convention on Persistent Organic Pollutants (UNEP POPs)

The Stockholm Convention is a global treaty to protect human health and the environment from persistent organic pollutants (POPs). POPs are chemicals that remain intact in the environment for long periods, become widely distributed geographically, accumulate in the fatty tissue of living organisms, and are considered to be toxic to humans and wildlife.

The twelve persistent organic pollutants are:
- Aldrin
- Chlordane
- Dieldrine
- Endrine
- Heptachlor
- Hexachlorobenzene
- Mirex
- Toxaphene
- DDT
- Polychlorinated biphenyls (PCBs)
- Polychlorinated dibenzo dioxins ("dioxins" or PCDDs)
- Polychlorinated dibenzo furans ("furans" or PCDFs)

The first nine are commercial pesticides. All of these are restricted as to continued production and use.

Except for some limited historical applications in such things as defoliants (such as Agent Orange), PCDDs and PCDFs were not usually manufactured as commercial products. Small quantities of dioxins and furans are in various waste streams from commercial processes and are also a component of incomplete combustion products for other chlorinated aromatic chemicals such as chlorinated benzenes and PCBs. Commercial preparations of "pure" PCBs were typically commercial mixtures of PCBs, frequently diluted with trichlorobenzenes and tetrachlorobenzene. Fire involving a transformer containing one of these commercial preparations would generate substantial quantities of dioxins and furans. As a general class of chemicals, PCDDs and PCDFs are among the most toxic of known, man-made chemicals.

Polychlorinated biphenyls exhibit many of the chemical traits of the other eleven chemicals listed. They are considered toxic by many regulating authorities, and there is a large volume of regulatory restrictions published concerning their manufacture, use, transportation, and disposal.

In 1997, United Nations Environment Program was authorized to convene an Intergovernmental Negotiating Committee (INC) to prepare an internationally binding agreement to control manufacture and use of POPs. The INC completed work on the Stockholm Convention on Persistent Organic Pollutants in December 2000. The Stockholm Convention was adopted in conference on May 22, 2001. Approximately 160 governments have signed the Stockholm Convention. Approximately 28 governments have submitted an article of ratification; the Stockholm Convention gains the force of law when the 50th government has submitted an article of ratification.

Parties to the Stockholm Convention are required to:

a. Eliminate use of PCBs in transformers, capacitors, and other equipment containing liquid PCBs by 2025, subject to the following priorities:

　　i. Make determined efforts to identify, label, and remove from use equipment containing greater than 10 percent PCBs (100,000 ppm) and volumes greater than 5 liters

　　ii. Make determined efforts to identify, label, and remove from use equipment containing greater than 0.05% PCBs (500 ppm) and volumes greater than 5 liters

iii. Endeavor to identify and remove from use equipment containing greater than 0.005% PCBs (50 ppm) and volumes greater than 0.05 liters

b. Promote the following measures to reduce exposures and risk to control the use of PCBs:

i. Use only in intact and non-leaking equipment and only in areas where the risk from environmental release can be minimized and quickly remedied
ii. Not use in equipment in areas associated with the production or processing of food or feed
iii. When used in populated areas, including schools and hospitals, take all reasonable measures to prevent fire from electrical failure and regularly inspect equipment for leaks

c. Ensure that equipment containing PCBs is not imported or exported except for purposes of environmentally sound waste management.
d. Except for maintenance and servicing operations, do not allow recovery for the purpose of reuse in other equipment of liquids with PCB content above 0.005% (50 ppm).
e. Make determined efforts designed to lead to environmentally sound waste management of liquids containing PCBs and equipment contaminated with PCBs having a PCB content above 0.005% (50 ppm) as soon as possible but no later than 2028.
f. Identify and manage other articles containing PCBs greater than 0.005% PCBs (50 ppm) such as cable sheaths, cured caulk, painted objects, etc..
g. Provide a report every five years on progress in eliminating PCBs.

Pure PCB Transformers

Continued operation of pure PCB transformers poses a risk to the owner of a catastrophic transformer failure which could spread PCB. The expense of disposing of a PCB transformer is minimal compared to the cost of remediating a PCB spill. Failed PCB transformers have been known to contaminate floors, walls, ceilings, drain systems, etc. A fire in one office building's basement involved a PCB transformer. When the fire reached the transformer, smoke from the fire spread PCB contamination throughout the entire, multi-story office building. Cleanup took years and cost millions of dollars. The moral of this and many other PCB disaster stories is this: you should consider replacing PCB units at the fastest rate your organization can afford.

Many countries do not yet have PCB laws and regulations that allow for in-country disposal. Additionally, some countries do not allow PCB owners to export their equipment outside their own country for disposal. Under these circumstances, the owner of PCB equipment may be tempted to dispose quickly and cheaply of contaminated equipment through unregulated channels. This is not advisable. Nobody wants to introduce PCB into the environment. Also, the experience in the United States has been that people who improperly disposed of PCB prior to the regulations have still been liable for cleanup after the regulations were written. It costs much more to dispose of PCB equipment through unqualified vendors and then pay to clean up their mess later than it does to dispose of it the right way in the first place.

Drained and flushed PCB transformer carcasses can be landfilled in some countries. The advantage to landfilling is its low cost. As a disadvantage, PCB owners disposing of equipment through landfilling must consider the future potential expense of being involved in a remediation project at the landfill. Several hazardous waste landfills have turned into environmental remediation sites with those who contributed waste being obligated to cover their portion of the expenses. Again, this second expense would be much greater than the cost for a more complete and permanent disposal method in the first place.

Several companies worldwide offer the service of incinerating entire PCB transformers. The advantage to this disposal method is that the transformer is completely destroyed. The disadvantages are that the process is expensive and that the ash and slag from the incinerator are still landfilled.

Many transformer owners prefer recycling PCB transformers over landfilling or incineration. In an environmental recycling process, the wood and paper of the transformer core is separated from its metal components. They are incinerated along with the liquid. The copper, steel, aluminum and brass components of the transformer are cleaned of PCB contamination and recycled. The advantage of recycling is that the former owner receives a certificate of destruction, thus releasing him from future liability for the equipment. The cost of recycling is usually between the costs for landfilling and for incineration.

Mineral Oil Transformers Containing PCB

Various countries regulate PCB contaminated mineral oil at different levels – some at 500ppm PCB, others at 50ppm, still others at 2ppm or even 1ppm. Each country's regulations are somewhat different than others, but one principle holds true for transformer disposal in each of them.

When disposing of a mineral oil transformer that contains PCB the primary concern is properly disposing of and controlling the oil itself. The metal of a transformer does not contain any. This has led many people to believe that a drained and flushed transformer can be disposed of and processed as scrap metal. However, the wood and paper of the core and coil assembly, as well as many of the metal surfaces, still contain or are covered with PCB-contaminated oil. If a dismantling facility handles these drained transformer carcasses in the same manner as they do other "scrap metal," they will inadvertently spread PCB contamination throughout their operation (as well as the facilities of those they send materials on to). Many transformer owners seeking the lowest cost method for disposal of drained and flushed carcasses of PCB-contaminated mineral oil transformers have contaminated scrap yards. Many of these same transformer owners have subsequently paid for environmental remediation of these same scrap yards. In the United States, for example, mineral oil for use in electrical equipment was unregulated below 50ppm PCB. Oil outside the transformer, however, was regulated at 2ppm. Many people sent transformers with oil above 2ppm and below 50ppm to non-environmental scrap operations. Once the oil from these transformers in drips and drops found its way onto the ground, it spread around the facility by foot and vehicle traffic, wind, rain, etc. Many of these facilities became environmental remediation sites, funded by the companies that sought low-cost transformer disposal.

The lesson to be learned: use only environmental recycling or disposal processes for electric equipment whose insulating fluids contain a regulated amount of PCB.

Mineral Oil Transformers Containing No PCB

Just when North American transformer owners thought their lives had become easier because they had removed nearly all the PCB from their electrical systems they have had to face a new environmental restriction. In the 1990's, mineral oil with no detectable levels of PCB became regulated as an environmental contaminant by virtue of being a hydrocarbon. Spills of non-PCB mineral oil still need to be treated as an environmental cleanup situation to prevent hydrocarbons from migrating into the water table.

The lesson learned from improper scrapping of mineral oil transformers containing some amount of PCB now applies to mineral oil transformers containing no PCB. They should not go to a scrapping facility that does not treat the mineral oil as an environmental hazard. Before sending scrap transformers to a recycling facility, the owner must conduct an environmental audit to be assured that the materials he sends will not be handled in a way that allows the mineral oil out of the processing area and into the environment. Without such assurances, the equipment owner may face legal and financial liabilities for contamination caused by the actions of the disposal firm.

Chapter 10

Windings

Purpose and Function

The transformation of electric current and voltage – a transformer's basic reason for existing – takes place in the windings. Two windings together - and nothing else – could form a transformer. It would be inefficient and awkward, but it would be a transformer.

A transformer's primary windings provide the medium for the flow of the alternating current that produces the flux that magnetizes the core steel. The secondary windings provide the medium for flow of the alternating current induced by the flux radiating from the magnetized core steel.

The ratio of the number of turns of conductor in the primary coil to the number of turns in the secondary coil determines the output voltage (and current) of the transformer.

A transformer with 100 turns on the primary winding and 100 turns on the secondary winding would have a ratio of 1:1. Such a transformer would put out the same voltage and current put into it.

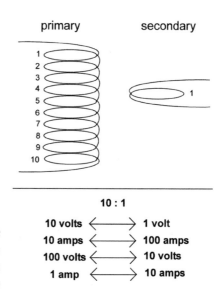

A transformer with 10 turns on the primary winding and 100 turns on the secondary winding would have a ratio of 1:10. This ratio would produce 10 times the voltage at 1/10 the amperage. With 100 volt, 100 amp input you would have 1000 volt and 10 amp output.

Reversing this ratio and putting 100 turns on the primary winding and 10 turns on the secondary winding would yield a 10:1 ratio. With 100 volt, 100 amp input you would have a 10 volt, 1000 amp output.

(None of these simple examples takes into account the core or other resistance losses.)

Basic AC Theory

- **Frequency and Sine Waves**
    - Frequency = 60 cycles/second = 60 hertz
    - Monitoring the electrical flow of AC energy can be demonstrated in time.
    - If we use a clock for measuring, we can asign time and position of the electrical flow.

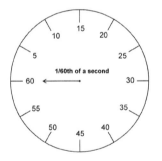
total measured time on clock = 1/60th of a second

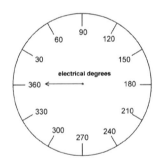
1/60th of a second as measured in electrical degrees

- The 360 electrical degrees represent 1/60th of a second of time, and the function of the electrical energy is flowing from + to -.
- We can make a graph of electrical flow in a time span of 1/60th of a second.

- Starting with the 0 or 360 degree we can plot this on the scale at the first point placing an X on its location on the graph.

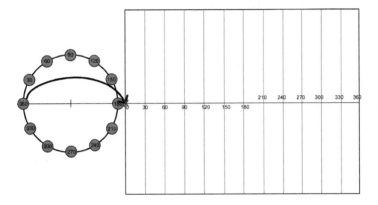

- We can plot the 30 degree point,

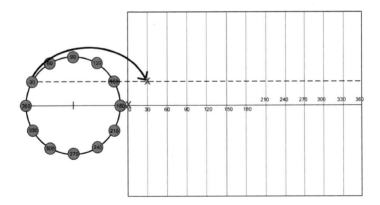

Windings | 337

- We can plot the 60 degree point, and 90 and so forth till all the degrees are plotted on the scale.

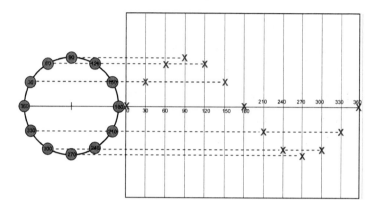

- By removing the lines, that placed the points, we can begin to see the curve of the graph.

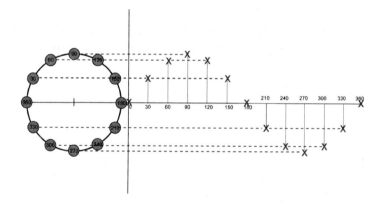

- Tracing the points we can see the curve of the "sine-wave."
- This illustration represents the positive and negative flow of electricity in a 60th of a single second.

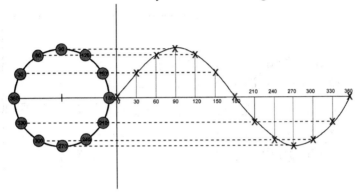

- **Conclusion (for a 60 hz system)**

    - The sine-wave represents the flow of electricity from the positive 60hz back through the neutral to the negative 60 hz.

    - This cycle occurrence happens 60 times in a single second going from positive to negative and back again.

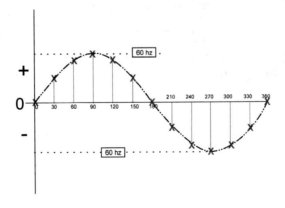

Single Phase Transformer Polarity

- Single phase transformer nameplate terminals are marked X_1, X_2 (low voltage), $H_1 H_2$ (high voltage).
- These terminals are connected to finish leads of windings which are placed over the core legs.
- By definition H_1 and X_1 have the same polarity (refering to the figure below) since the finish leads of the windings exit around the core in the same direction.
- The instantaneous induced voltage at these terminals (H_1 and X_1) has the same electrical sign and this is referred to as "Subtractive Polarity."

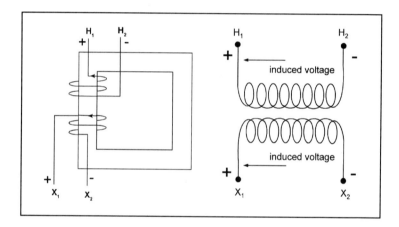

- If H_1 and X_1 exit around the core in opposite directions then this is referred to as "Additive Polarity"
- ANSI/IEEE Standard C57.1200 states that single phase transformers in sizes 200 KVA and below having high-voltage ratings 8660 volts and below shall have "additive polarity." All other single phase transformers shall be "subtractive polarity."
- Polarity is of no importance for a single phase transformer application using only one unit.
- If single phase transformers are placed in parallel or connected for three phase operation, polarity is very important.

Three Phase Transformer Polarity

Phase Sequence and Angular Displacement

- Polarity phase by phase is no different in a three phase transformer from what it is in a single phase transformer.
- Since there are various ways to connect the windings of each phase, polarity alone does not fully describe the phase relation between the high and low voltage systems.
- Phase sequence (phase relation) is determined by the winding polarity, the winding connections and terminal designations. These items are shown as electrical vectors on the nameplate. All vectors rotate counter clockwise to produce the phase sequence measured in electrical degrees.
- Angular displacement is the number of electrical degrees that exist between the high voltage and low voltage sinewaves for various winding connections.

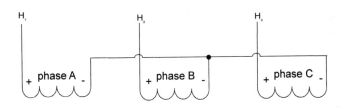

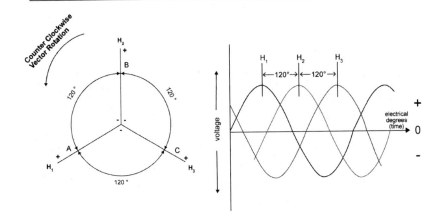

Phase Sequence and Angular Displacement of a WYE-WYE Connection

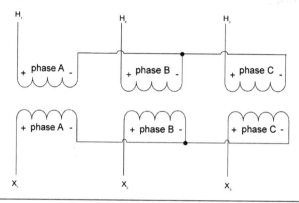

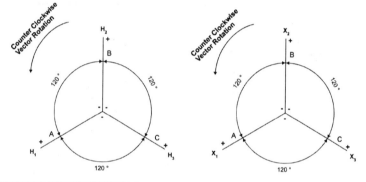

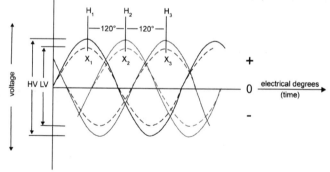

- Electrical "Sine-Waves" of these voltages
- Phase Sequence of $H_1H_2H_3$, $X_1X_2X_3$ voltages
- Angular Displacement of 0 electrical degrees between high voltage and low voltage

Phase Sequence and Angular Displacement of a WYE-DELTA Connection

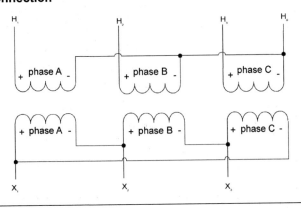

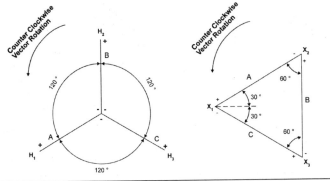

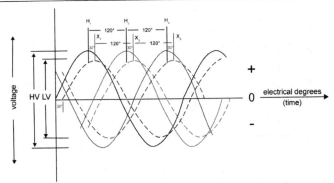

- Electrical "Sine-Waves" of these voltages
- Phase Sequence of $H_1H_2H_3$, $X_1X_2X_3$ voltages
- Angular Displacement of 30 electrical degrees between high voltage and low voltage

Types

Design Criteria

A transformer's designers and builders need to balance the following considerations for a transformer's windings:

Minimum resistance from
- Materials
- Eddy Currents
- Connections

Dielectric strength
- For ordinary operation
- Against various voltage stresses, such as lightning or switching surges.

Heat transfer
- From the winding materials to the liquid insulation
- For dry transformers, to the air

Mechanical strength
- For the forces of ordinary operation
- To withstand short-circuit forces

All at a Minimum Cost!

In some situations, the cost of the materials, (copper, steel, paper and oil) is the first thing considered, rather than the efficiency and regulation. Depending upon the size of the unit, this can be the deciding factor. The cost of copper and steel seem to change daily. With that in mind, the number of turns used in the transformer design of last week may not be the same this week. The designer may adjust the turns ratio according to the price of the materials.

Materials

A transformer winding is made from one of two materials: copper or aluminum.

The primary advantages of copper coils are:

- Greater mechanical strength
- Higher electrical conductivity (which results in smaller coils)

The primary advantages of aluminum coils are:

- Price
- Efficient heat dissipation for sheet wound (at smaller capacities)
- Lighter weight coils

Comparison of Physical Properties of Aluminum and Copper

Property	Aluminum	Copper
Electrical Conductivity at 20°C Annealed	62%	100%
Weight per cubic inch at 20°C	1.56 oz.	4.8 oz.
Specific Heat cal/(g·°C)	0.21	0.003
Melting Point	660°C	1083°C
Thermal Conductivity 20°C cal/cm²	0.57	0.941
Mechanical Strength Annealed 1000psi	13	32
2500 kVA transformer with 44kV HV	13,900 lbs*	14,700 lbs*
*Total transformer weight.		

Table 10.1

Conductor Shapes

Winding conductors come in one of three forms: round wire, rectangular conductor, or sheets

Design engineers size conductors (whether wire, rectangular or sheet) according to the amount of current they need to carry. Undersized conductors cause higher losses through the friction (too many electrons trying to move through too small of an area). This friction (resistance) transforms energy that previously had a potential to be useful, into heat. Heat represents not only wasted energy (which equals wasted money), but also accelerates the aging process within the transformer — which we discuss in the section 'The Unique Effect of Elevated Temperature on Aging' on page 406.

As current carrying requirements increase, designers must increase the cross section of the conductor. Round wire can be used in the high voltage coils of distribution units where the current requirement is fairly low. The more current to be carried, the larger the wire must be. At a certain cross-sectional dimension, using rectangular rather than round conductor minimizes eddy current losses. At another cross-sectional dimension, eddy current losses are minimized with multiple-strand conductor.

Wire and rectangular conductor impedes the flow of electricity less than does sheet material.

Even in well-designed transformers, some percent of current will be lost to resistance and eddy currents as it flows through the coils. These losses produce heat, which the windings must be designed to dissipate. Paper or wood spacers between turns or layers provide passageway or ducts for the heated insulating fluid close to the conductor to rise and be replaced by cooler fluid* from the lower part of the tank.

---

*Or air, in an air-cooled, dry-type transformer.

Testing and Monitoring

The characteristics of winding conductors that can be measured in the field – winding resistance and transformer turns ratio - were determined at the factory during manufacturing. They should not change over the life of the transformer (winding insulation is covered under the next major heading, "Solid Insulation"). If either of these measurements does change, there is a problem with the transformer (see Electrical Testing, part 3, p. 187).

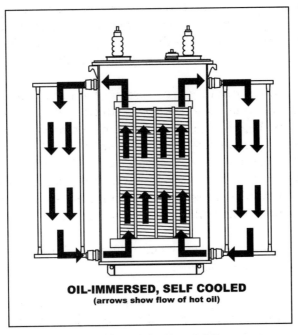

Figure 10.1 - Oil-Immersed, Self-Cooled diagram

Chapter 11

The Core

Purpose and Function

The core captures and consolidates the electromagnetic field produced by the primary winding, and directs this field to where it can most efficiently induce current in the secondary winding. Cores were made of iron (or, today, steel) because this material carries and confines the magnetic flow much better than air.

Brief History - Why Shell Form? Why Core Form?

In the early 1900's a machine called a transformer was beginning to be developed to transmit electrical power to factories, businesses, homes, etc. from power plants where electricity was generated. Westinghouse Electric Corporation decided in this time period that all power transformers produced in their factories would be designed and constructed as shell form. General Electric, also entering into the manufacturing of transformers, decided that all of their power transformers would be designed and constructed as core form.

Since it is more economical to transmit electrical power at higher voltages while using larger kVA sized transformers, both Westinghouse and General Electric continued to developed their respective shell form and core form technologies to the present standard of 765 kV and approximately 1200 MVA, 3 Phase.

In the late 1940's Westinghouse Electric decided that using shell form construction on power transformers with voltages to approximately 138 kV and 50,000 kVA was not as economical as core form construction. They developed and introduced a core form power transformer production line in their Sharon, Pennsylvania manufacturing facility for smaller and medium sized power transformers. Westinghouse continued to use shell form construction for higher voltages and larger kVA sizes until they (ABB) closed the Muncie, Indiana facility in 1996.

General Electric never used shell form construction while producing larger sized kVA transformers up to 765 kV. The G.E. facility in Pittsfield, Massachusetts closed in 1986. Their core form extra high voltage (EHV) technology was purchased by ABB for use as an alternate construction to shell form in the Muncie, Indiana plant.

U.S. manufacturers Allis-Chalmers and McGraw Edison (later Cooper Industries) used shell form construction exclusively for their higher voltage, larger kVA sized transformers.

Allis-Chalmers originally used a round coil shell form construction (1940 - 1970), and then abandoned this construction in 1970 for the more popular rectangle coil style used by Westinghouse, McGraw Edison, and Cooper Industries.

Notable foreign manufacturers of shell form transformers, include: IEM, Mitsubishi, Hyosung, Jeumont Schneider, Eficek, and ABB, (Cordoba).

The Magnetic Circuit

Electromagnetic Induction

When an alternating voltage (e) in the form of a sine wave is applied to a winding on an iron core, a very small alternating current (i) will flow (see Figure 11.1). This small alternating current (i) will cause magnetic flux to flow in the core. This small current is

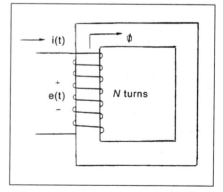

Figure 11.1

called the **exciting current**. The magnetic flux will assume an alternating sine wave shape similar to the applied voltage. The magnetic flux is almost entirely contained in the core since the magnetic **reluctance** (resistance) is much lower in core steel than oil or air (see Figure 11.2).

If another winding is then placed on the core an alternating voltage will be induced across it by the flow of magnetic flux. This process is called **electromagnetic induction** (see Figure 11.3).

Magnetostriction

The sine-wave represent the flow of electricity from the positive 60 Hz back through the neutral to the negative 60 Hz. This cycle occurs 60 times in a single second going from the positive to negative and back again. This is the basics of the process we call **magnetostriction**. Here is the transformer without any electrical flow (Figure 11.4).

As the electrical current moves as indicated by our sine-wave, the magnetized core does as well through the magnetic flux. The core pulls inward at the peak of the positive side of the cycle (Figure 11.5).

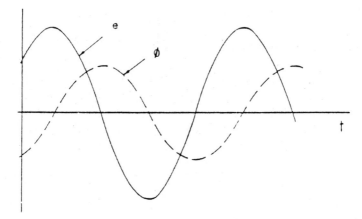

Figure 11.2

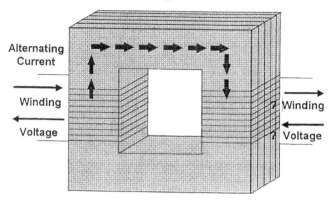

Figure 11.3

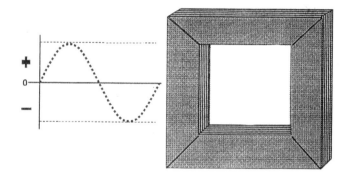

Figure 11.4

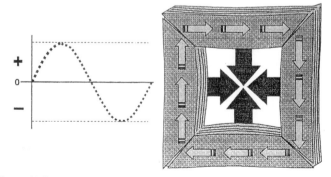

Figure 11.5

When the electrical flow again reaches the 0 point on the sine-wave, the transformer relaxes (Figure 11.6).

As the electrical current moves to the negative point of 60 Hz on the sine-wave, the magnetized core again pulls inward because of the magnetic flux (Figure 11.7).

When the electrical flow again reaches the 0 point on the sine-wave, the transformer once more relaxes (Figure 11.8).

The very slight contraction and expansion of the core steel laminations, measured in parts per million (micro-inches per inch of steel length), occurring 120 times a second, during magnetization and demagnetization is called **magnetostriction**.

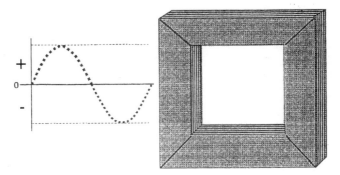

Figure 11.6

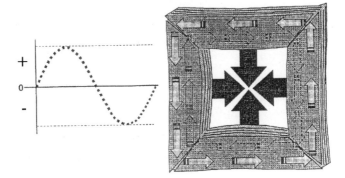

Figure 11.7

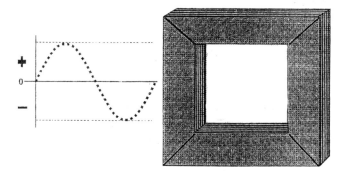

Figure 11.8

Core Noise

The **magnetostriction** or movement of the core back and forth 120 times per second causes vibration of the core steel and thus noise. This noise is heard as a core hum when energized by AC voltage. For 60 Hz transformers, the core sound is generated at 120 Hz because the core respond to both the positive and negative AC voltage wave peaks for each cycle. As shown in Figure 11.9, the magnetostriction properties of the core are directly proportional to the flux density (B) for which the core is designed. As the flux density in the core is increased, the elongation and hence the contraction in length of the core steel laminations is effected. The higher the core flux density, the higher the noise level of the core. The core noise level is composed of two components. The frequency (pitch) as measured in cycles per second and the intensity (loudness) which refers to the amplitude as measured in decibels (dB) of sound pressure level.

Core items that effect its noise level:
- Flux density (B) is the major factor
- Core geometry (core form or shell form)
- Core dimensions
- Core building quality

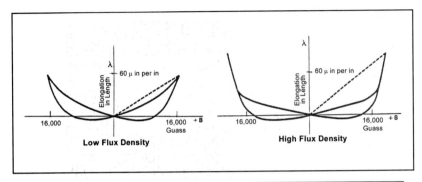

Figure 11.9

Total Transformer Noise (combination of noises)
- Core noise level
- Internal lead structure vibrations
- Core supports and coil restraining scheme
- Tank vibrations/resonance
- Oil pumps
- Cooling fans
- Load tap changers
- The total transformer noise level as measured before transformer shipment, per ANSI/IEEE standards, is designed not to exceed noise levels listed in the standard.

Three Phase Core Form Core Construction

A typical 3 phase core is shown in Figure 11.10. The three phase magnetic flux similar to Figure 11.2 flows in the core legs as φ_1, φ_2 and φ_3. These three phase fluxes are 120 electrical degrees apart. They add vectorially to a resultant of zero magnitude as indicated during transformer operation.

The physical construction of the core is shown in Figure 11.11. The core is constructed with thousands of thin laminations of gain-oriented steel that serves as the path for the magnetic flux (φ_1, φ_2 and φ_3).

The core legs are the vertical members over which the winding phases are placed. The core yokes are the horizontal members (top and bottom) that bridge the core legs and complete the magnetic circuit.

The core joints are a critical transition point in the core construction where the core leg laminations meet the core yoke laminations. The laminations are mitered (usually at 45 degrees) at this transition area for easier passage of the magnetic flux. To further reduce the reluctance (resistance) of the core joints to flux passage, the laminations are typically step lapped as shown in figure 11.12. Effective control of core gaps (1/8" to 3/16" maximum) produces lower core loss, lower sound levels and reduced core hot spots.

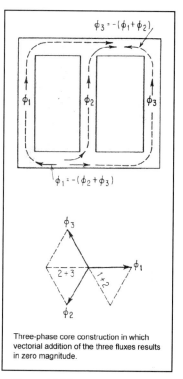

Three-phase core construction in which vectorial addition of the three fluxes results in zero magnitude.

Figure 11.10

The cross sectional area of the core leg is composed of various widths of steel laminations to accommodate cylindrical phase windings used in power transformers (See Figure 11.11).

Core ducts, usually high density pressboard material are placed at intervals in the core stack height to provide oil ventilation as shown in Figure 11.11.

An intentional core ground (thin copper strap, 2" wide) is inserted between core laminations and ground. This procedure keeps the core from assuming a dangerous potential above ground.

Core (No Load) Losses

The normal core loss of a transformer is comprised of two components, eddy loss and hysteresis loss.

Eddy Loss - when the magnetic flux in a core lamination changes going through its sine wave configuration, a voltage is induced in the lamination. The current flowing in response to this voltage is called an eddy current. The eddy loss due to this current is approximately proportional to the square of the thickness of the lamination. Each core steel lamination is treated with a chemical coating (carlyte) to insulate the laminations from each other. This process increases the interlaminar resistance of the core stack and discourages eddy loss between laminations. Typical grades of core steel range from M-6 (14 mil thick), M-4 (11 mil thick), M-3 (9 mil thick). The thinner the core lamination, the lower the core loss component due to eddy loss. One can imagine the eddy loss value of a core that was solid with no laminations. As one might also expect, the thinner the lamination the greater the cost.

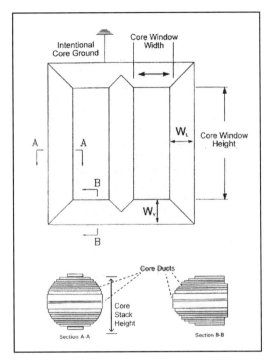

Figure 11.11

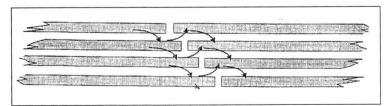

Figure 11.12 - Laminate Stacking of Core Steel

Hysteresis Loss - This part of the core loss occurs as the core steel magnetizes and demagnetizes from the application of the AC voltage. Silicon alloy core steel was developed especially for transformer steel laminations. The silicon alloy content (approximately 3%) reduces the resistance to the magnetization effects of the steel and prevents increased losses.

The laminations are cold rolled and specifically annealed to orient the grains or iron crystals to obtain very high permeability and low hysteresis to the magnetic flux.

Conclusion (Core Loss)

- Energy losses in the core steel are caused by the magnetic flux flowing in the core.
- The magnetic flux is generated by a very small alternating current (i) called the exciting current which flows in the winding when an alternating voltage is applied.
- The core loss (eddy and hysteresis) and the exciting current increase as the flux density (B) increases in the core (see Figure 11.13).
- Some transformers are energized, causing core loss, without being loaded for long periods of time. This can be a significant expense for the power company. An average evaluation by them for this expense is $3000 to $3500 per kilowatt of loss.

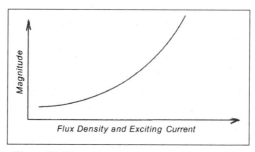

Figure 11.13

History

The cores first used in early experiments were simple bars of iron. All of the principles behind a transformer's operation apply in a transformer designed and built around a bar of iron, but this would not be a very efficient transformer. A straight iron bar core initially would cost less to manufacture than do modern transformer cores, but the additional electricity costs to operate such a transformer would soon eat up the savings. Other materials capture, intensify and direct the electromagnetic field more efficiently than iron does; as well as core designs other than a straight bar.

Materials

The ideal core material would produce no friction between magnetic molecular particles as the AC magnetic field continually reverses its direction (minimizing hysteresis loss). To minimize eddy currents, the core must be constructed of laminations as thin as possible. (The thickness of the laminations cannot be reduced beyond a certain point because the laminations would become too weak mechanically).

In the early 1900's, thin, flat, soft iron laminations were used as the core material. Then it was found that small quantities of silicon (Si) alloyed with low carbon content minimized the frictional loss of hysteresis.

Until the mid-1930's, iron-silicon alloy core-lamination sheets were produced from slabs by successive rollings at red heat (1370 °C followed by annealing at approximately 900 °C). This process produced sheets in which the metal's crystals were nearly randomly oriented. In 1935, N.P. Goss discovered that cold-rolling and high temperature annealing tended to orient the steel's molecular crystals in the same direction. This core material further reduced hysteresis losses since the most efficient magnetization occurs when the magnetic lines of force flow parallel to the steel grain (high permeability and low hysteresis loss). Today, this grain-oriented grade of steel containing about 3% silicone is used almost exclusively on distribution and large power transformers since core losses are a significant factor in the overall cost of transformer operation.

Insulating adjacent sheets of laminations from each other also minimizes eddy current losses in the core materials. Without insulation the laminations would act as one block of steel, rather than as individual laminations, with a block's higher eddy current losses.

Energy Efficient Transformers

The choice in the transformer core steel is critical for today's transformer core losses. It is important that good quality magnetic steel be used. The choice will be based on how you evaluate total ownership costs (toc), which include, no-load (core), load (coil) and auxiliary losses. Substation transformers usually have fans and pumps that allow them to operate at higher capacities. The loading can take place in one, two or three stages and there is a corresponding auxiliary loss for each stage that must be considered during loss evaluation. Calculating the value of no-load and load losses for substation units is similar to calculating losses for distribution transformers, but there are some major differences. The cost of the distribution system is not included in the capacity cost for a substation transformer. The cost of load losses in a substation transformer must be calculated for the various stages of cooling and loading. Auxiliary losses must be included in substation transformer losses and in some cases the cost of preventive maintenance tasks are included in the toc. It is very difficult to calculate accurately the core losses over the lifetime of a transformer, and users employ many different formulas.

Since new low-loss transformer technologies such as amorphous metal cores, cryogenics, and laser-etched silicon steel became available; most government agencies have encouraged their use. Utilities widely buy energy efficient distribution transformers, which contribute significantly to the utilities system losses because

Transformer losses on a typical utility system[1]

Transformer type	1-million kWA	
	Core	Coil
Generator step-up	18	89
Power substation	67	138
Distribution substation	97	114
Distribution	328	127
Total	**510**	**468**

of their large numbers. But substation transformers, because of their large kVA size, contribute more per unit to system efficiency than distribution transformers.[1]

All transformer manufacturers these days use core steel that provides low losses. To achieve this, high permeability cold rolled grain oriented silicon steel is used. The permeability of core steel is a measure of how many times better it is at conducting magnetic flux then air.

Another material that warrants consideration is Allied-Signal's Metglas™. Metglas™ is the trademark name for amorphous metallic alloy technology utilized primarily in magnetic applications ranging from transformer cores to miniature electronic cores. The ability to form Metglas™ in very thin sheets (.03mm thick, 200mm wide) results in low eddy current losses. Also the physical properties (Fe 78%, B 13%, Si 9%) result in very low hysteresis losses.

Properties	**3% SiFe**	**Metglas™ 26055-2**
Thickness (mm)	0.3	0.03
Max. working temp °C	650	150
Space factor (%)	95-98	80
Resistivity (μΩcm)	45-48	137
Saturation induction (T)	2.03	1.56
Loss at 1.3T/50Hz (W/kg)	0.64	0.11
Loss at 1.5T/50Hz (W/kg)	0.83	0.27

Comparison of properties of commercial iron-based amorphous material and grain-oriented silicon steel.[2]

[1] Kennedy, B, *Selecting Energy Efficient Substation Transformers* (1999)
[2] Moses, AJ, *Factors affecting economic use of amorphous metal transformer cores*

Amorphous metal is made in a process that eliminates the crystalline formation found in conventional metal. The material is sprayed unto a fast-moving chilled substrate, and rapid solidification at around 10^6 °C/s ensures that the strip so formed does not contain grains like a normal metal but has a glassy or amorphous state at room temperature.[2] Metglas™ is sometimes referred to as glassy metal because of the similarity with glass in atomic structure. The efficiency of the material is a direct result of the manufacturing process and materials.

In the 1950's and 60's, the demand for power increased rapidly and because there was a limited availability of distribution transformers, purchase was largely based on availability. Later, load growth began to fall, demand for new transformers dropped, and price and quality became major factors. In the 1970's, increased energy costs made us concerned about total cost of ownership, and the concept of loss evaluation was born. The cost of ownership includes: purchase price, cost of operation and losses over a 30 year period. Today, about 2% of total electricity generated is used up as transformer core losses; therefore, a large portion of high-cost generation is wasted.[2]

It is well know that the no-load loss of a typical amorphous core is only about 30% of that of a high-efficiency silicon core.[2] In 1992 it was estimated the U.S. would have a 125GW deficit by 2000 and a 230GW deficit by 2005.[3]

Amorphous vs. Silicon Transformer Core Loss Comparison (watts) [3]

Rating (kVA)	In-Service Silicon	Hi-Efficiency Silicone	Amorphous
100	320	160	54
1000	2400	1200	420
2500	4800	2400	850

[2] Moses, AJ, *Factors affecting economic use of amorphous metal transformer cores*
[3] Nagel, WD, *Ultra efficient amorphous core transformers improve distribution system efficiency* (1992)

Eventual renewal of the U.S. electric utility and industrial distribution transformer systems with amorphous transformer will reduce utility core losses by 7000MW and industrials by 3000MW. In addition to eliminating need for generating capacity, continuous savings would be realized in energy consumption and related environmental effects:

Annual kWh Savings: 85 Billion - The equivalent of running all U.S. generating plants for 10 days!

Equivalent Annual Energy Savings[3]:

Oil   140 million tons
Coal   42 million tons

Avoided Environmental Emissions[3] (coal):

SO_2 2 million tons
CO_2 80 million tons

But these numbers need to be put into context. For instance, in 1990 the cost of no-load losses was estimated between $1/W and $10/W.[2] Utilities with surplus power will assign a lower dollar amount to losses and utilities routinely at peak demand will assign a higher amount to the losses. Distribution transformer core losses are a burden on the generation of the most expensive electricity. This is because the losses are supplied continuously so consequently, a large portion of the high-cost generation is wasted.

When amorphous material was first introduced in the late 1970s it cost around 100 times that of silicon steel.[2] Today's higher amorphous transformer manufacturing cost can result in transformer prices of 20% to 30% higher than silicon steel transformer alternatives.[3] The amorphous metals are cheaper to produce consuming about 20% of the energy required to produce silicon steel.[2] The current problems still seen to be in the handling and construction of the cores. The material is brittle and very stress-sensitive. The material becomes hard and brittle after heat-treating to optimize its magnetic properties so it is essential that nothing be done in assembly to introduce stress. Cores made of amorphous material of any given rating, tend to be larger than those made of silicon steel. This makes it difficult to build the traditional stacked cores. The manufacturer tried bonding several layers of the ribbon together in hopes of being able to use it in the making of stacked cores, but the product was too difficult to use and was withdrawn from the market. In addition this process degraded the space factor in comparison to a silicon steel stacked core. The cores of most transformers using amorphous metal are wound from the ribbon. This leaves basically two choices, either the core is cut to allow the coils to be put into place, or the coils must be wound directly around the core.

Cost vs. Benefit

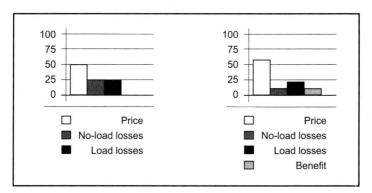

Figure 11.14 Silicon Core Steel   *Figure 11.15 Amorphous Core steel*

In the long run, improving transformer efficiency is often more cost effective than adding new generation, or demand side management programs.

In October 1992, President George Bush signed the Energy Policy Act. The Energy Act called for a study to determine if a transformer standard would be technologically feasible, economically justified, and results in significant energy savings. Preliminary analyses indicate that such a standard would improve efficiencies in both dry-type transformers and oil-filled distribution transformers. In response to the governmental activity, the National Electrical Manufacturers Association

(NEMA) approved NEMA Standard TP-1-1996 "Guide for Developing Energy Efficiencies for Distribution Transformers." This standard cites minimal efficiency for each transformer kVA rating and could become the basis for a standard if such legislation occurs.[4]

Figure 11.16 - Core Form - being assembled

Figure 11.17 - Core Form - fully assembled

[4] Eilert, P, *Dry-type Transformers, Codes and Standards Enhancement (CASE) Study* (2000)

Figure 11.18 - Mitered corner construction

Figure 11.19 - Shell Form

Chapter 12

Solid Insulation

Stresses on Solid Insulation

A transformer in-service is exposed to a variety of stresses - requiring both dielectric and short-circuit or mechanical strength.

- Dielectric Stress
- Short-Circuit Stress
- 50/60 Hertz Excitation
- Lightning Impulses
- Switching Surges
- Short-Circuit Through Faults

Dielectric Stresses

It is a fact that a transformer operates under 50/60 Hertz excitation. The resulting 100/120 cycle mechanical vibration of the entire structure under the influence of 50/60 Hertz power results in voltage stress. This stress is continuous and may be as high as 105 or 110% of rated voltage, depending on the tap range and the limits imposed by ANSI standards.

A second major dielectric stress involves lightning surges, arising from direct lightning "hits" to the electrical system or from indirect surges in the vicinity of lightning strikes. The result is a traveling wave on the transmission line which may "hit" the terminals of a transformer (Figure 12.1).

The third stress consideration involves switching surges. These impulses may be caused by a number of system operations or abnormalities but are often due to line switching.

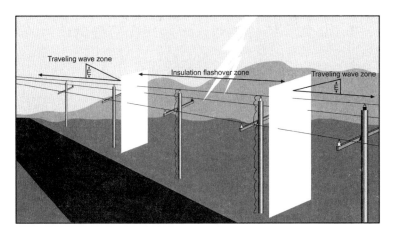

Figure 12.1 - Lightning striking an electrical transmission line consisting of two parts - structures experience flashover and adjacent travelling wave zones where insulation flashover does not occur.

The dielectric stresses are defined in terms of time in microseconds. Each stress presents greatly differing voltage-time relations (Figure 12.2). It would take a long graph to show these relationships in complete form. The essential parts, however, are shown by looking at the two ends of such a plot. For example, in the case of lightning surges, these are represented in laboratory studies as a wave (an undesirable disturbance which reaches its peak or crest value at 1.2 microseconds and decays to half value in 50 microseconds). The typical switching surge used reaches maximum value in 230 microseconds and decays to half value in 2000 microseconds. In contrast, the 60 Hertz time requires 4167 microseconds just to reach its crest value.

Short-Circuit Stresses

In time past, lightning was the primary stress. In recent years, through-faults (that is, an excessive current flowing through a transformer) not only have been more severe, but they have also been more numerous.

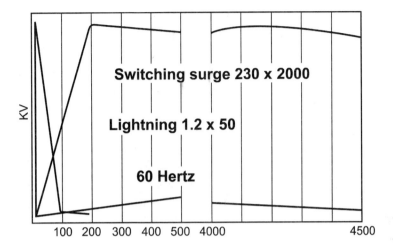

Figure 12.2 - An in-service transformer must be designed to withstand various dielectric stresses. Different voltage-time relationships are required for each of these three stresses.

The primary cause of transformer failure as the result of a short circuit is not because of direct thermal damage of the insulation but rather the mechanical forces produced in the windings. As the large kVA sizes of transformers have increased in recent years, so have the mechanical forces that are stressing the integrity of their insulation systems. The interaction between windings results in both horizontal and vertical electromagnetic forces.

The vertical or axial force causes the low-voltage and the high-voltage windings to shift with respect to each other, a condition called "telescoping" (Figure 12.3). These forces make the windings take positions that will increase the magnetic flux of the system. If two windings are in series, the electromagnetic force varies as the square of current. For example, a short-circuit current 20 times normal will produce 20^2 (or 400 times) the normal stress.

The vertical force between primary and secondary windings results because it is impossible to exactly balance the low- and high-voltage electrical center lines. The only way the interaction of the two windings could be eliminated would be to have the coils

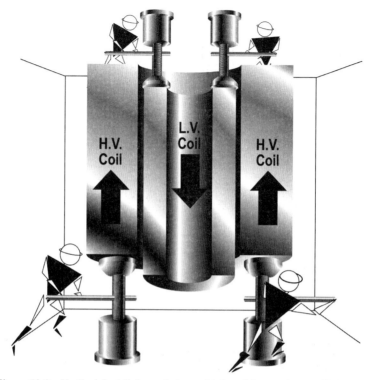

Figure 12.3 - Vertical (axial) forces between high and low- voltage coils in core form transformers in a through short-circuit.

occupy the same space at the same time — an impossibility! Therefore, the resultant force between the two coils can be minimized only when the manufacturer is diligent in his design and workmanship. Because of the complexity of this problem in manufacturing, no two transformers are alike, each having its own characteristics. Although the horizontal force (radial or hoop) is the major force (Figure 12.4), it is the vertical component that is most difficult for which to design and manufacture.

Unfortunately, all three stresses — dielectric, mechanical and thermal — often occur simultaneously.

A lightning strike can create a fault near a transformer when it is loaded and hot, just as well as when it is relatively unloaded and cool. Indeed, in a severe lightning storm it may be overloaded

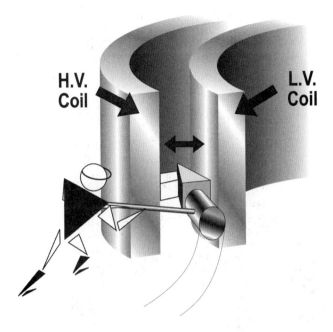

Figure 12.4 - Horizontal repulsion (radial) force between high-voltage and low-voltage coils in core form transformer in a through short-circuit.

since other equipment may be (already) out of service. Then the overheated insulation system may be subjected to a lightning strike, followed almost at the same time by severe mechanical shock with accompanying heat from the high-fault current.

This fault current can cause a hot spot temperature of 500 °F due to the I^2R losses. If the transformer does not fail, but continues to operate, this temperature will reduce dramatically inasmuch as the rapid rise in hot spot temperature is not accompanied by a corresponding Btu input.

This is the situation, therefore, under which a transformer must operate through proper design, manufacture, quality control testing, and preventive maintenance. Rectifier and furnace transformers operate on "short-circuit current" or are subject to many short-circuit operations and, therefore, require greater mechanical strength to resist deformation from short circuits.

60 Hertz Normal Excitation Voltage

The principal concern is both the major and phase-to-phase insulation. Failure in-service (or in a test) can result from:

- Inadequate insulation through error in design or manufacture.
- Improper installation.
- The presence of partial discharges; as the term implies, this is a partial breakdown of the insulation structure within a transformer, sometimes referred to as "corona," but the proper terminology is "partial discharge." Such discharges are undesirable as progressive tracking or formation of gas may result.
- Long-time deterioration as the result of thermal and voltage stress.
- Improper or no maintenance.
- Excessive loading and/or overloading.

60 Hertz Transient Over Voltage

Normal operation for three-phase transformers results in a phase-to-ground stress that is 58% of the phase-to-phase stress. However, this relation is altered substantially under single-phase fault conditions. The stress on the major insulation may be higher than normal by 30% for a grounded system or 73% for an ungrounded system. This transient stress should normally be no problem, but in the case of design or manufacturing error, it is possible that insulation puncture, once initiated, would persist after returning to normal operating conditions.

Lightning Impulses

Failure during a lightning impulse (no more than 50 microseconds) involves an outright puncture or tracking of the insulation. This point of failure may occur anywhere in or on the impulsed winding or in another winding because of the oil stress produced by the voltage oscillations. The very rapid increase and collapse of voltage principally stresses the bushing, winding leads and the first few turns of the winding. If the puncture is through major insulation of phase-to-phase insulation, failure is instantaneous, whereas if the puncture is in the minor insulation, final failure may be caused by the resultant 60 Hertz follow-current.

Refer again to Figure 12.2. Note that a non-linear, initial voltage distribution followed by a series of oscillations occurs until a linear voltage is achieved. This places considerable stress on the minor insulation.

It is such considerations that illustrate insulation coordination - the correlation of the various discharge characteristics of protective devices and basic insulation levels.

Switching Surges

In contrast to the lightning strike, the front of this wave is slow enough so that the voltage distribution is approximately linear. The impulse is transferred to other winding terminals approximately in proportion to the turns ratio between the two winding. This means that any resultant failure will most likely occur in the major or phase-to-phase insulation (Figure 12.5 - top).

Within any one phase, two other critical areas exist:

- Surge can creep around the major insulation and cause failure (Figure 12.5 - bottom)
- The high voltage can puncture the primary layer insulation (Figure 12.6)

This is why furnace transformers require extra insulation between turns to guard against high voltage caused by arcing.

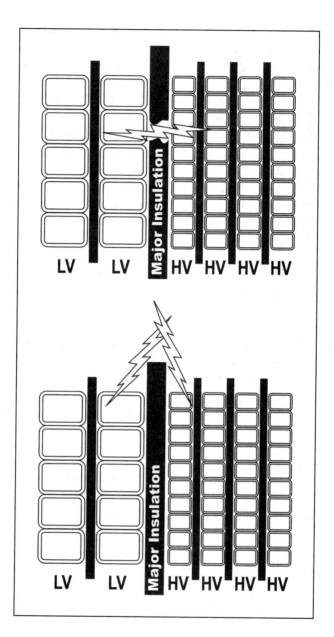

Figure 12.5 - (Top) - Failure due to puncture of major insulaton between high and low-voltage windings. (Bottom) - Failure due to creepage around major insulation.

Short-Circuit Through-Faults

A well-built transformer has been likened to a tough prize fighter who is hit hard and often—yet manages to come back for more. The "hits" a unit absorbs are in the form of short circuits. For distribution transformers, approximately 450 "hits" occur during a normal lifetime, or as many as 54 through-faults of varying magnitudes in one year! Power units are subjected to even more numerous and more severe faults.

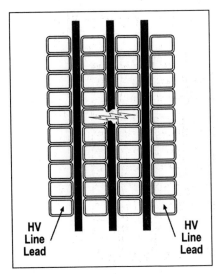

Figure 12.6 - Failure due to puncture of primary layer.

By their very nature, transformers seem compelled to shake themselves apart or tear themselves to pieces. The more common vertical or axial forces (Figure 12.2), operating on the high- and low-voltage windings, tend to pull one another apart; that is, telescope. Figure 12.7 illustrates how the major and minor insulation are affected.

The horizontal, repulsive radial forces tend to make the low-voltage coil buckle inward and the high-voltage coil buckle outward (Figure 12.4). The major insulation would be primarily affected, while the minor insulation only secondarily.

Solid Insulation Life

How long will a transformer last? How long should it last? How long could it last?

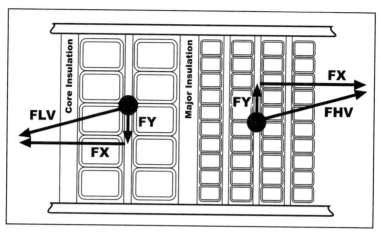

Figure 12.7 - Fault Movement

Theoretically, the materials in your transformer should be able to last for three or four hundred years. We don't see that happening though.

A study of transformer life conducted in 1965 showed an average life expectancy of 22 years. A study conducted in 1988 showed an average life expectancy of 11 years! These studies do seem to indicate that older, over-built transformers last longer than newer, smaller transformers built with less insulation and tighter tolerances.

When your transformer does fail, chances are that it will fail in the windings due to weakened solid insulation; 85% of transformers fail this way! Solid insulation is the Achilles heel of transformers. If you properly maintain your transformer's solid insulation you could get as many as 50+ years of reliable operating life.

Transformers reach the end of reliable operating life before they finally fail and stop operating. Reliable operating life is defined as the time when a transformer can withstand the stresses of normal operation. As a transformer ages, its insulation loses strength until finally it falls outside of the design criteria for its particular application. As a transformer's solid insulation loses strength and becomes less able to withstand the stresses of operation, it becomes less reliable. At some point it becomes unreliable enough that it should be replaced in a scheduled, planned manner rather than waiting for it to fail and cause an unplanned power outage.

The mechanical strength of cellulose insulating material is measured either by its tensile strength or its degree of polymerization (DP).

Tensile strength measures the physical strength of a sample of insulation. DP measures the length of cellulose molecules, which relates directly to the insulation's physical strength. The longer the insulation's molecule chains (polymers), the stronger the insulation is, and vice versa.

Brand new Kraft insulation has a DP of about 1200 (17,000 psi tensile strength). After the manufacturing process wraps the insulation around the conductors, winds the conductors around the coil and bakes the whole thing in a dry-out oven, the DP comes down to about 1000 (14,200 psi tensile strength). Over the life of the transformer different conditions cause the insulation to depolymerize. As the insulation's DP gets close to 200 DP (3400 psi tensile strength) it is no longer reliable. According to the IEEE Std.C57.91-1995, *IEEE Guide for Loading Mineral-Oil-Immersed Transformers*, the end of reliable insulation life is defined as a DP value of 200 or 75% loss of tensile strength (using 14,200 as a base, this would be 3400 psi). Some transformers fail before reaching 200 DP and some keep operating beyond 200 DP, but transformers in the 200 DP range have definitely reached the end of their reliable operating lives. The owner of such a transformer should be making immediate plans for replacement.

Conditions That Destroy Solid Insulation

What causes a transformer, with a theoretical life of three or four hundred years, to reach a DP of 200 within ten or twenty (or even fifty) years?

Many forces are at work in an operating transformer to hasten the day it sees 200 DP. Chief among these forces are:

- Water
- Oxygen
- Oil Oxidation By-Products
- Heat

Water

How does water affect solid insulation in transformers?

The Kraft paper used for transformer insulation is similar to the brown paper used in grocery bags. Imagine setting your paper bag of soup cans and a dozen eggs down in a water puddle, then picking it up five minutes later. What do you think will have happened to the paper's tensile strength?

Operating experience with transformer insulation over many years has shown that moisture in microscopic amounts – not gallons – is the cause of more electrical breakdowns than any other impurity.

Recognition of the importance of extremely small amounts of moisture has grown immeasurably with the increase in voltage stress and reduced BILs. This truth was first published in a 1902 AIEE paper.

Indeed, moisture "constitutes a hazard not only to the dielectric performance of the oil itself but also to insulations that are immersed in the oil." Thus, we should all agree that water is "enemy number one." What is not known or generally accepted is that a water problem may begin before ASTM standard dielectric test D 877 reveals it. The late F. M. Clark has also stated that:

> "The electrical user...recognized the dielectric hazard which is presented by the presence of water

but rarely does he recognize the equally disastrous effect which traces of moisture have on the long-time usefulness of the insulated equipment, even when the amount of moisture is not sufficient to cause dielectric difficulties."

Factory and Installation

It may take up to two years to build a large transformer and during that time the unit may collect up to 10% M/DW of the insulation. Therefore, initial drying is required prior to assembly and ultimately shipping.

Consensus opinion says that the paper insulation of a new transformer leaving the factory must contain less than 0.5% water by dry weight. Thus, a certain amount of water is always present and can be readily accommodated by the transformer. Therefore, the purchaser of the new equipment should not tolerate moisture contents higher than 0.5%.

Indeed, it is recognized that a transformer will perform to all its design criteria and nameplate rating with a moisture content as high as 1.0% by weight. Nevertheless, this higher water content does affect the longevity of the transformer. An analogous situation would be like adding a 50 pound weight on the back of a world class marathon runner. When he drops out of the race at five miles and can't finish the twenty-six mile, 385 yard race, the reason is obvious - too much handicap! It's the same with 1.0% moisture content in a transformer. It won't finish the race.

Figure 12.8 illustrates how critical small quantities of moisture are to the dielectric strength of transformer mineral oil, as compared with silicone and askarel (polychlorinated biphenyl or PCB).

Table 12.1 shows that the water solution in the oil even for 20 ppm is still less than one-tenth the residual water in the paper for vary large (EHV) power transformers.

To minimize the entrance of moisture during transit from the manufacturer, a new transformer tank is filled with dry gas (such as nitrogen) under pressure. Moisture content should remain essentially the same as at the time the unit was shipped. If the gas pressure and oxygen content are not the prescribed values, high probability exists that the transformer has been contaminated with atmospheric air or moisture. In such a case, as part of the installation procedure, a dry-out becomes necessary.

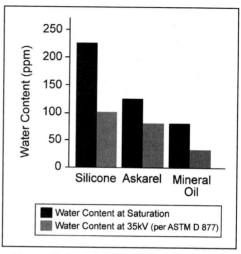

Figure 12.8 - Solubility of water in various transformer fluids.

When a large power transformer leaves the factory, hopefully it has been dried sufficiently so that the residual moisture is typically 0.5% or less. Nonetheless, moisture can enter into the insulating system in these three ways:

1. External Sources

    - Opening the transformer tank during installation or maintenance which results in exposing the unit to the atmosphere.
    - Accidental leaks in the tank — around lids, bushings, or inspection cover gaskets.

2. Partly External Sources

    - The precipitation of water from the atmosphere on inner tank surfaces (lid and side walls in the air space).
    - Moisture condensation on the tank, dripping back on a terminal board, tap changer, leads, or through the oil onto and into the cellulosic paper insulation.

Comparison of Water in Oil and in Cellulose Insulation

EHV Power Transformer	Cellulose Insulation (pounds)	Oil (gallons)	Residual Water in Cellulose Insulation		Water in Oil (ppm)	
			0.1% (gallons)	0.2% (gallons)	5 ppm (gallons)	20 ppm (gallons)
A	11,700	10,000	1.4	2.8	0.05	0.18
B	33,000	15,000	3.95	7.9	0.07	0.27
C	44,000	20,000	5.27	10.55	0.09	0.36

Table 12.1

- Temperature changes in transit plus leaks or loss of nitrogen because cellulose is an excellent desiccant (moisture absorber).

3. Internal Reactions

As the insulation system thermally ages, water is produced. Nothing can prevent this gradual deterioration process:

- The decomposition of the cellulose — the aging of insulation resulting from heat — as the combination of hydrogen and hydroxyl (OH) groups.
- As one product of oil degradation — the chemical reaction of the oxygen present with hydrocarbons.

Several laboratory studies have shown that when cellulose degrades, moisture is produced. An applicable everyday analogy is what happens to old newspapers used as a shelf-liner — embrittlement occurs after several weeks, even at room temperature.

Degradation may be expressed in terms of the degree of polymerization. Recalling that the deterioration rate of the paper depends mainly on the temperature, Figures 12.9 and 12.10 then give a picture of what happens when heat liberates water form the cellulose. Clark, Fabre, and Pichon made this startling observation — water liberated from the paper accelerates the rate of cel-

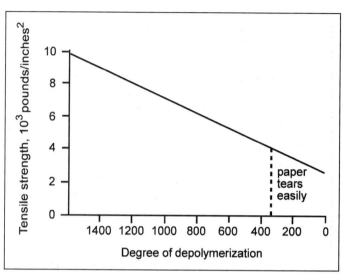

Figure 12.9 - Relationship between tensile strength and degree of depolymerization.

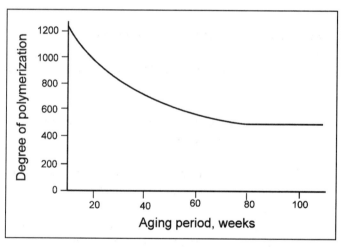

Figure 12.10 - Aging of paper in model transformers at 85 °C top oil temperature (paper at about 100 °C)

lulosic degradation, otherwise dependent solely on temperature. Analysis has thus shown that water content may fall in a range between 0.5 and 2.5%.

When oil oxidizes, it also produces water as a degradation product together with acids, sludge, and other polar compounds. The relative proportions of these products depend on the oil and the aging conditions. If it is supposed that a typical transformer contained ten tons of paper and pressboard and 8000 gallons of oil, then in oxidizing to 0.5 mg KOH/gram the oil would produce between 2 and 20 gallons of water. This quantity would increase the water content of the paper between 0.1 and 1.0% by dry weight. Therefore, the minimum quantity of moisture from the oil and paper together totals between 0.6 and 3.5%.

The significance of this compounded effect is that at 2% total water content the rate of aging is between 6 and 16 times and at 4%, 12 and 45 times that at the initial 0.3% water content. As a typical example, during a twenty-five year period the moisture accumulation may be 3.3% by dry weight in the insulation.

Characteristics of Water Affecting Insulation

What are some of the general characteristics of water in oil or paper that may affect the life of transformer insulation? Keep these in mind:

1. Water is always present in commercial insulations. In view of the sources this is to be expected.

2. Water (with air a standard of comparison) is a polar liquid having a high permittivity or dielectric constant, varying between 79.5 and 81 (pure water only). By contrast, the dielectric constants of most commercial insulations range from 2.0 to 7.0. Therefore, water is attracted to areas of

strong electric field. It is not uniformly distributed throughout the insulating system of the transformer, but in fact may concentrate in the most dangerous parts of the system.

3. Water is strongly electropositive and attracted to negatively charged electrodes and repelled from positively charged electrodes.

4. Water's electrical conductivity is largely due to the presence of polar contaminants, such as salts and acids.

5. Water is a universal solvent except for fats and sulfur. This solvent property of water can be used to advantage. The freshly formed oxidation products are water soluble rather than oil soluble. To what extent? Better than 10 to 1, so that water removed will tend to wash out these fresh, acidic, volatile materials.

6. Water-soluble acids produced by oxidation of the oil will combine with water left to assist or promote corrosion to exposed metal parts in transformers; for example, on the underside of manhole covers. Rusting also occurs on the metallic parts of a transformer, even though covered with oil. Thus, it is of paramount importance that water be removed regularly.

7. Water is a catalyst (a substance that accelerates a chemical reaction without itself being consumed) for almost all reactions. Nearly all reactions require at least a small amount of water.

8. Water forms corrosive nitrous and nitric acids in combination with the gases released by the ionization of air during partial discharges.

9. Oxidized oils contain water-soluble polar groups which orient themselves on a water surface so the hydrophilic

(water loving) group is in contact with the water. The result of this interaction weakens the strength of the film at the interface of the water and oil. The reduction in the interfacial tension (IFT) of a transformer oil is an indication of the early stages of the oxidation of an oil.

10. The cellulose has greater affinity for water than oil (hundreds of times). Water will replace the oil in oil-impregnated cellulose (moisture migration). Thus, the cellulose is sucking up the moisture because cellulose is hygroscopic while the oil is getting rid of the water because it is hydrophobic; that is, the oil repels water.

Oil and Water Do Mix!

A familiar cliché says, "Oil and water do not mix." Undoubtedly, you have heard it many times but... "it ain't necessarily so!" It utterly fails to be true in the engineering sense. Oil and water do mix and in a number of ways. Moisture can exist in transformers in a number of different forms, including: free water, suspended water, dissolved water, and "chemically bound water."

1. Free Water

> Free water is generally present in many shipments of new, clean insulating oil, whether in drums or tank cars. It is also frequently found at the bottom of transformer tanks and is relatively harmless at this point. The water can be seen with the naked eye when drawing a sample, provided the sample valve is located at the exact bottom of the unit.
>
> While the transformer is clean and has a high IFT (30-50 dynes/cm), water entering the unit will show up very rapidly in the bottom of the transformer because the water is heavier than the oil. If the free water did not have access to the insulating system of the transformer it would be relatively harmless in this location (Murphy's Law says that

some free water will end up in the cellulose. This signals the onset of an inevitable moisture problem.). While it is true that the oil in immediate contact with the free water on the bottom of the transformer will be water saturated at the lower temperature, it still is relatively harmless to the transformer. This remains true so long as the water goes down the sides of a unit rather than splashes over the top of a transformer's core and coils.

2. Water in Suspension (Emulsified)

Moisture in oil free from solids in suspension is not a serious problem. When oil is new and free from contaminants, water will quickly settle to the bottom of the tank.

The problem of moisture becomes significant in the presence of cellulosic fibers and polar contaminants dissolved in the insulating liquid. When contaminants including oxidation products, fibers, or other debris are present in oil, water tends to cling to such contaminants or be absorbed by them. This water in suspension is closely associated with oil decay products. Under this circumstance, water and water-soluble acids will combine with acids and sludges in an oxidized transformer oil.

Water in suspension cannot be seen with the naked eye and is not detected with a flat disc ASTM D 877 dielectric test.

If an oil is oxidized to any extent, any water coming into the transformer will partially be absorbed into the oil decay products. As the decay products build up in the oil, the surface tension of the water of the interfacial tension between the water and the oil is lowered dramatically. These oil decay products align themselves at the water-oil interface, reach into the water, and draw the water into existing decay products; they are thus hydrophilic (love of water) and will continue to suck the moisture. This heavier decay molecule will then find its way into the cellulosic insulation, or into areas of high electrical intensity, thus causing a reduction in the insulation resistance. Here is another effect of Murphy's Law: the water-saturated oil decay molecule has a preference for the coolest part of the transformer (bottom and fins) and areas of highest electrical stress. Aged oil holds much more water in solution than new oil, and hot oil more than cold; therefore, hot, aged oil contains more water than cold, new oil.

The necessity to identify this problem very early in the history of the transformer is most apparent. Running a power factor test on the oil is one method of detection. A high power factor test in transformer oil will indicate a problem, and one of the causes could be water trapped in oil decay products - that is, water in suspension.

Consider now the most degrading form of moisture - water in solution or dissolved water.

3. Water in Solution (Dissolved Water)

   a. The established fact that oils could actually dissolve moisture has still not been completely recognized, even though this fact has been established for over 60 years.

   F.M. Clark stated that the presence of dissolved water in oil "constitutes a hazard not only to the oil itself but also the cellulose immersed in the oil." The amount of water that may be dissolved is primarily a function of the temperature and the condition of the oil. This reminds us that the primary conditions which affect solid insulation are temperature and moisture.

   It might prove interesting to study Figure 12.11 to follow the moisture in cellulosic insulation under

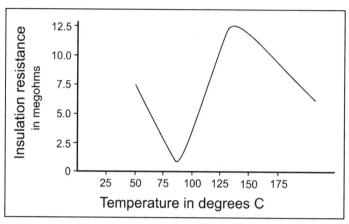

Figure 12.11 - Classic relationship - fibrous materials

heated conditions. It shows the variation of the insulation resistance with temperature. The increased resistance of a given path through the dielectric at lower temperature is exaggerated by the presence of free moisture.

If a piece of insulation is gradually heated and the moisture is free to escape, the resistance decreases at first, because of the moisture being vaporized and again increases because of the vapor being expelled at the higher temperatures. After the expulsion of the moisture the material begins to change in nature and ultimately begins to carbonize and behaves as a high-resistance conductor. The resistance then falls rapidly to a very low value. While this curve is typical, the exact temperature at which the minimum and maximum resistance values occur depends on the area of the electrodes, the amount of moisture content, the facility with which it may escape and the chemical and physical changes that occur due to the temperature.

Relationships between the dissolved moisture content and each of the associated media — cellulose, oil and air — are normally stated in terms of temperature and parts per million (ppm). An analogy may assist in trying to picture in your mind the quantity in ppm of dissolved water in a given insulating medium: one part in a million is comparable to one minute in 1.9 years, or one hour in 114 years!

b. Water saturation in transformer oil - A significant relationship between new oil and dissolved water has been found to exist. The plot in Figure 12.12 is called the "Cloud Point Curve." Three general statements can be made regarding this curve:

- All oils dissolve more moisture at higher temperatures than lower temperatures.
- Cooling the unit gives oil the opportunity to unload its water.
- The greater the extent of refining operation, the less the solubility for water.

The curve shows "soluble" water on the right of the curve and "free" water to the left, respectively. As an example, the curve shows that at 40 °C, oil has the ability to hold 120 ppm of water "in solution." If the temperature of the oil with 120 ppm of water is raised to 60 °C, the oil would still be sparkling clear, and a standard dielectric (D 877) conducted at this temperature would give a high reading somewhere in the range of 40 to 50 kV. However, if the temperature drops to 40 °C and lower, the oil would go from a "clear sparkling" oil to a "cloudy" oil. Therefore, the point at which the transformer oil goes from the soluble water state (clear oil) to the free water state (cloudy oil) is called the "cloud point" (ASTM Test Method D 1524); ergo, Cloud Point Curve!

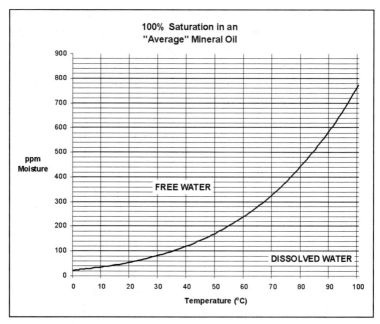

Figure 12.12 - Effect of temperature on solubility of water in new transformer oil ("Cloud Point Curve").

4. Hot Transformer Oil Holds More Moisture

   Let's consider a transformer with free water in the bottom. The layer of oil next to the free water will be saturated with water. As the transformer heats up due to the load (especially above 40 °C), so will the "water-saturated oil" heat up and, at the new temperature, the oil will become unsaturated. This oil with water in solution will rise up through the transformer past the cellulosic insulation and eventually flow into the cooling fins.

5. Cooling the Transformer

   As this oil and water combination comes down through the fins and cools, the water will precipitate out, recondense on the sides of the fins, and run down to the bottom of the transformer again.

It is apparent if a hot transformer oil contained an appreciable amount of water in solution, this water would have no (appreciable) effect on the overall power factor of the transformer (in solution) at the time of test. However, if the transformer was tested later, after the oil had cooled off, the overall power factor might be higher.

This is due to water phasing out of solution and being absorbed by the insulation. Free water brought on by cooling of the oil will end up in the insulation and also as water in suspension. The power factor of the oil and of the insulation will be seriously affected by the moisture. Inasmuch as decay products of oil have an affinity for the coolest part of the transformer and the decay products also attract water, the inevitable result is internal rusting that eventually results in leaks. Therefore, 95% of leaks in transformer cooling fins are from the inside of the fins to the outside. This statement would not necessarily hold true, however, when a transformer is located in an area of severe chemical fallout.

6. The Effect of Oil Refinement

The higher the degree of refining, the less water can be held in solution for any given temperature. Generally, a more viscous oil dissolves more moisture than a less viscous oil; for example, lubricating oils dissolve more moisture than insulating oils.

Chemically Bound Water
It is a little know fact that a very small amount of water is a virtue in order to maintain the mechanical strength of cellulose. The cellulose contains a certain amount of "chemically bound" or absorbed water in the OH and H groups of the glucose molecule which may be released at high temperatures. The present day trend toward transformer operation beyond nameplate ratings makes it necessary to keep this fact in mind. As far back as the early 1940's

the possibility was recognized that the beneficial "chemically bound" water would be released through excessive temperature and then lose its virtue altogether. This "chemically bound" water ends up as free water or water in solution.

Nonetheless, some moisture is required in the environment during the manufacturing of coils; if the humidity is too low, cracking and breaking of some materials may result.

Dissolved Water and Insulation Failure
 1. Equilibrium and Imminent Danger

> Consider now the distribution of dissolved water between the insulating liquid and the cellulose, particularly at the point of equilibrium. This occurs in two substances. In a transformer, these two materials are the transformer oil and cellulose insulation. The point of equilibrium is reached when the moisture content remains constant for at least twelve hours. No further change in quantity or distribution of the water takes place, regardless of the humidity or the number of days of exposure at a given temperature.
>
> Figure 12.13 shows what happens to the water as final equilibrium develops between various insulating liquids and air. F. M. Clark found that the three fluids exhibit a water solubility relationship, which is proportional to the relative humidity of its environment.
>
> The absorption of moisture by the cellulose insulation involves two stages:
>
> First stage — The rate at which mineral oil absorbs moisture from the surrounding atmosphere depends on the intimacy of contact between the oil and the atmosphere. Figure 12.14 again show a proportional relationship between water content and exposure of transformer oil to humid air under these conditions:

Volume of liquid	7500 cc
Surface area exposed	375 sq cm
Depth of liquid	20 cm
Temperature of exposure	25 °C

Second stage—When water is present in the oil or the cellulose insulation of a transformer at the relatively low temperature of 25 °C, it tends to reach a common equilibrium, that is, moist oil gives up water to the dry cellulose (Figure 12.15) or wet cellulose gives up water to dry oil (Figure 12.16).

In a sealed container, once the equilibrium condition is reached between the water content of the cellulose and that of the insulating liquid, this equilibrium is not easily changed.

In this context, please note that in the manufacture of electrical equipment, this point of equilibrium is rarely if ever achieved. These experiments of F.M. Clark and others were controlled life tests under laboratory conditions.

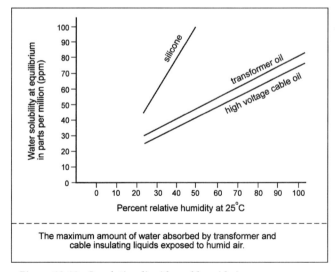

Figure 12.13 - Insulating liquids and humid air.

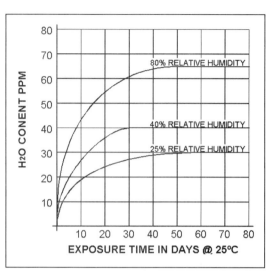

Figure 12.14 - The absorption of water by mineral transformer oil.

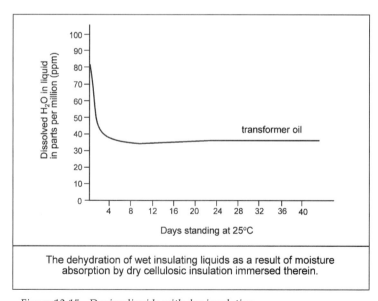

Figure 12.15 - Drying liquids with dry insulation

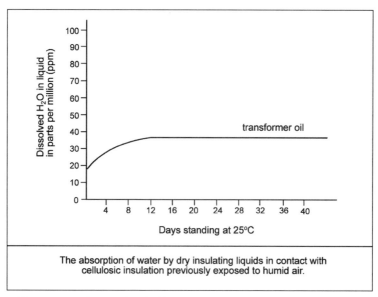

Figure 12.16 - Water absorbed by oil

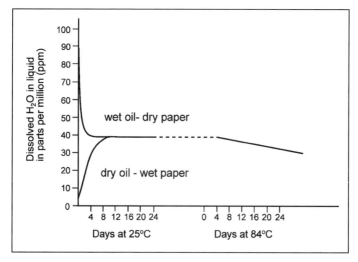

Figure 12.17 - The characteristics of the moisture equilibrium set up between mineral transformer oil and the impregnated cellulose insulation (1940).

We should note also that Clark reaffirmed the validity of these earlier tests (1940) in 1962. "A dynamic situation is the order of the day. For example, in the case where temperature is increasing, the system moisture transfers from paper to oil, and when the temperature drops the system moisture transfers from oil to paper." Again, we recognize the significance of the cloud point condition. Clark also found that equilibrium remains substantially unchanged when each set-up is heated from 25 °C to 84 °C liquid temperature.

All this has been said in order to focus attention on the danger (Figure 12.17) which results when moisture is allowed to be dissolved by the insulating liquid. Thus, Dr. Clark speaking of the problems involved in drying out wet transformer insulation, concludes:

"When once the moisture is absorbed by the impregnated cellulose, unless it is present in large amounts, it is released again only by drastic treatment, which involves the application of temperatures in the range of 95 °C to 100 °C and higher."

Recall the limiting temperature for Class 105 impregnated cellulose is 100 °C - 105 °C. Clark's data was obtained using kraft materials immersed in oil.

2. Equilibrium and Vapor Pressure

An equilibrium situation for the paper and oil system can also be expressed as a condition in which the water vapor pressure in the paper is equal to that in the oil at any given temperature.

The pressure of the saturated vapor in equilibrium with a particular medium is the water vapor pressure in that medium. This water vapor pressure expressed in millimeters of mercury (or Torr), should be kept at a minimum. Normal atmospheric pressure is 14.7 pounds per square inch absolute, equal to 29.92 inches mercury absolute, and equal to 760 mm mercury absolute. The classic reference for moisture content of kraft insulation as a function of temperature is the "Piper Chart," see Figure 7.2, page 279.

The use of inert gas, such as nitrogen in enclosed transformers, has become the standard. This gas is used to minimize the exposure of the oil and paper from the moisture and oxygen of the air. An equilibrium condition develops between the moisture content of the fibrous materials and that in the gas space.

In this regard, a modified Piper equilibrium chart has proved useful for predicting and interpreting the behavior of the moisture within transformers containing cellulosic material and a gas space. In using the chart, keep in mind that the vapor pressure is not the same as the pressure of the gas, but is only the partial pressure of the water vapor in the gas.

3. Ionization and Dielectric Breakdown

   Ignoring dissolved moisture opens the door to additional chemical degradation. Such deterioration of the cellulose results in the formation of additional moisture and acidic polar materials. The presence of these materials leads to ionization under electric stress.

   Not all the deterioration which leads to the failure of insulation can thus be ascribed to contamination. The second broad category involves ionization or partial discharge within the dielectric structure. (Figure 12.18)

Thinking in terms of the cellulosic atomic structure, the freeing of one electron from a stable atom or molecule constitutes ionization. This usually refers to the progressive ionization involving a large number of electrons.

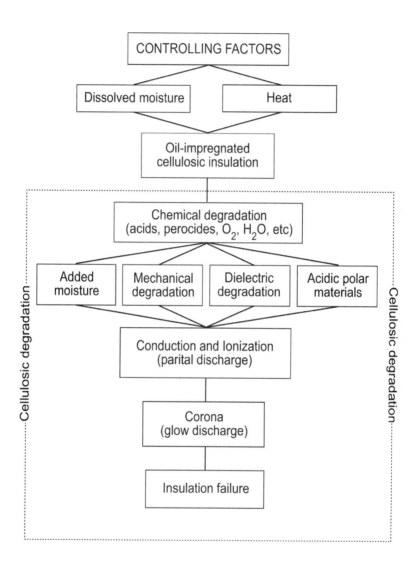

Figure 12.18 - The progression toward insulation failure.

Ionization does not necessarily require any form of electrical discharge whatsoever. Nevertheless, as the voltage stress is further increased, that is, as the number of free electrons continue to be released, and electron avalanche begins. Dielectric degradation already underway leads to more moisture evolution, dielectric shrinkage, and gas formation, such as carbon dioxide and carbon monoxide within the dielectric structure. This results in unequal dielectric stress distribution with greater concentration of stress within the void. The cellulosic air space becomes "ionized," and at some point the ionization phenomena merges into a destructive type of a glow discharge, which is universally recognized by the term "partial discharge" or "corona." This process damages many insulating materials, evolves gas such as ozone, and ultimately leads to dielectric breakdown.

All this has been said to demonstrate the truth that for continued transformer operation, cellulosic insulation must be maintained in a dry condition.

Oxygen

Free oxygen available to cellulose insulation will combine with the chain molecules in ways that break apart and shorten them, therefore shortening the life of the insulation (Figure 12.19). However, the greater impact on insulation comes from by-products of oil oxidation.

Oil oxidizes much more dramatically than cellulose, producing dozens of oxidation by-products. This process changes the appearance and characteristics

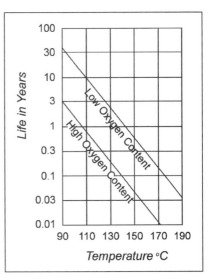

Figure 12.19 - Life of cellulose. Insulation vs. temperature. End of life defined as DP=200 (degree of polymerization). W. Lampe et al., "Continuous Purification and Supervison of Transformer Insulation Systems in Service."

of the oil. Cellulose insulation absorbs the by-products of insulating oil oxidation just as the paper filter element in an automobile's engine oil filter absorbs impurities from the motor oil. The impurities absorbed by cellulose insulation from oxidized liquid insulation break down the polymer chains in the cellulose, weakening it and shortening the transformer's life.

The Mechanics of Sludge Formation

An ASTM study has concluded that the process of oil oxidation consists of two main cycles of reaction:

1. The formation of oil-soluble decay products such as acids, beginning as soon as the oil is put in operation.

2. The change of the oil-soluble oxidation products into oil-insoluble compounds.

Sludge, precipitated first on the cold parts and then onto the hot parts of the transformer, will continue to oxidize and eventually become insoluble. Sludging takes place periodically rather than continuously.

Figure 12.20 illustrates the cross-section of a transformer coil with a resulting build-up of sludge by progressive layers. The layer in contact with the coil is relatively thin, and successive layers are progressively thicker. Now assuming that five successive layers of sludge have developed over a period of time, the first layer next to the coil, being thin and having good thermal contact with the heat source, will continue to oxidize and eventually become insoluble. As we progress from layer to layer, we will find varying degrees of hardness depending upon the extent of oxidation that has continued to take place since the sludge was laid down.

Oxidation can be controlled but not eliminated. Oxygen comes from the atmosphere or is liberated from the cellulose as a result of heat. Oxidation of the cellulose is accelerated by the presence

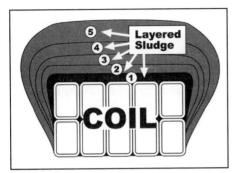

Figure 12.20 - Shows cross-section of transformer coil with five successive layers of sludge build-up. Layer one has already solidified and is now a "permanent part" of the transformer.

of certain oil decay products called "polar compound," such as acids, peroxides and water.

The Kraft paper, being extremely porous (probably 85 - 95% air), will absorb (or soak up) oil up to about 10% of the transformer oil fill. Therefore, the process of cellulosic degradation begins immediately, starting at the initial oil fill of the unit. In addition, both paper and pressboard have pronounced selective adsorption for the volatile organic acids formed as the initial oil decay products.

The first decay products - peroxides and water soluble and highly volatile acids - are immediately adsorbed by the cellulose insulation up to its saturation level. In the presence of oxygen and water, these "seeds of destruction" have a potent destructive effect on the cellulosic structure. This deterioration may begin long before any oil test data indicates a problem. The acids of low molecular weight are most intensively adsorbed by the cellulosic insulation in the initial period; later, the rate of this process slows down. The oxidation reaction may attack the cellulose molecule in one or more of its molecular linkages. The end result of such chemical changes is the development of more polar groups and the formation of still more water.

Heat

Nine years of data concerning the relationship of tensile strength of paper, aged in oil and in air, provided the basis for what has become know as "the 8 °C rule." The thermal life of class 105 insulation is halved for each increase of 8 °C or conversely doubled for each decrease of 8 °C.

The rationale for this guidline is found in the so-called "10 °C rule" which has been widely applied to evaluate the rate of chemical reactions.

By this it is meant that the rate of a specific chemical reaction can be expected to double for each 10 °C increment of temperature increase, all other factors remaining constant.

In the late 1940's, T.W. Dakin and G. Malmlow pointed out that aging is the result of one or more chemical reactions and that the rate at which chemical reactions proceed varies with temperature. Dakin reasoned that the relation of insulation deterioration to changes in time and temperature is therefore assumed to follow an adaptation of the Arrhenius chemical reaction rate law. This theory states that the logarithm of insulation life is a function of the reciprocal of absolute temperature.

$$\text{Log}_{10}\text{Life (hours)} = A + \frac{B}{T}$$

T = absolute temperature in degrees Kelvin (HST +273); HST= the hottest spot temperature in degrees Celsius.

A and B = Constants determined by the selected stresses imposed on the insulation (A) and by the material used in the insulation system (B).

Additional experiments have shown that this relationship of life and temperature has proven acceptable for many types of insulation structures up to 140 °C. Dakin's use of the Arrhenius chemical rate phenomena is still generally accepted as the best available basis for thermal aging.

Various investigators today, nevertheless, still do not agree on the length of life at any given temperature. While the 8 °C rule is well known, research now has evidence that heat may even be more critical and tends toward the 7 °C rule.

Thermal Evaluation Testing

Insulation Material Evaluation

The first procedure involves the relative life of a specific material in a particular environment. As an illustration, one concept of measuring insulation life used the relationship between temperature and loss of tensile strength. Another uses the relationship between temperature and loss of degree of polymerization (DP).

Even though numerous studies have been made, the results are fairly consistent, considering the possible variations in the technique. Table 12.2 from the IEEE Loading Guide shows us the following:

First, Table 12.2 projects how long it takes for paper to lose 50% of its tensile strength and also 75% of its tensile strength. But of what practical value is this? As an example, life at 110 °C is approximately 7.42 years when 50% of strength is lost and 15.41 years for 75% loss of tensile strength. If a 200 DP value is the end of life criteria, the time would be 17.12 years.

Table 12.2 gives an idea as to the time limit at which transformers may be safely operated at various temperatures and the equivalent loss of life; for example, when the life at 110°C is 180,000 hours, then the life at 200 °C is 104 hours, and in one percent of this time, or 1.4 hours, one percent of the life would be lost.

1. Insulation System Evaluation

    In this procedure small models of complete apparatus are used to evaluate the compatibility of various insulation materials.

2. Functional Transformer Testing

    Both insulating material and system evaluation testing have proven useful, but, only simulated tests of "full-sized" equipment can provide thermal endurance comparisons

End of Life Criteria	Normal Insulation Life	
	Hours	Years
50% retained tensile strength of insulation (former IEEE std C57.92-1981 criterion)	65,000	7.42
25% retained tensile strength of insulation	135,000	15.41
200 retained degree of polymerization in insulation	150,000	17.12
Interpretation of distribution Transformer functional life test data (former IEEE std C57.91-1981)	180,000	20.55

Table 12.2 - Estimate life hours based on various end of life criteria. From IEEE Std. C57.91-1995. Copyright 1995 IEEE. All rights reserved.

of various insulation systems and form the basis of guides for loading in-service filled transformers.

One accepted correlation of insulation life versus temperature for power transformers is shown in Figure 12.21. This table shows the corelation between hot spot temperature and aging acceleration factor.

The Unique Effect of Elevated Temperature on Aging

In spite of the complex scope of thermal degradation, its unique effect apart from mechanical stress had not been fully realized until recent years.

The increasing practice of recurring overloading of transformers beyond the nameplate rating, as well as higher average loads is due to improved technology and business economics. The effect of thermal aging appears to have even greater significance for larger MVA units. As a consequence, the role of thermal aging has finally been recognized as a prominent factor in life expectancy.

Loading Guide For Oil-Immersed Transformers

As a guide for predicting transformer aging and to recommend loading practices, loading guides were established. Early loading guides (1930's, 1940's) used mechanical deterioration alone as end of life criteria. More recent loading guides recognize that transformer life is influenced by thermal, mechanical and electrical stresses as well.

In 1995, the Institute of Electrical and Electronics Engineers (IEEE) combined previous guides, C57.91, C57.92 and C57.115 into one document: IEEE Std. C57.91-1995 *IEEE Guide for Loading Mineral Oil-Immersed Transformers*. This guide deals with the general recommendations for loading 65 °C rise mineral-oil-immersed distribution and power transformers. Recommendations for 55 °C rise transformers are included in the annex because a substantial number of these units are still in service.

In the previous guides, different insulation aging curves were used for power transformers and distribution transformers. The distribution transformer aging curves were based on aging tests of

Hot Spot Temp °C	Aging Rate	Percent Loss of Life[a]						
		0.0133[b]	0.02	0.05	0.1	0.2	0.3	0.4
110	1.00	24.00	-	-	-	-	-	-
120	2.71	8.86	13.3	-	-	-	-	-
130	6.98	3.44	5.1	12.9	-	-	-	-
140	17.2	1.39	2.1	5.2	10.5	20.9	-	-
150	40.6	0.59	0.89	2.2	4.4	8.8	13.3	17.7
160	92.1	0.26	0.39	0.98	1.96	3.9	5.9	7.8
170	201.2	0.12	0.18	0.45	0.89	1.8	2.7	3.6
180	424.9	0.06	0.08	0.21	0.42	0.84	1.27	1.7
190	868.8	0.028	0.04	0.10	0.21	0.41	0.83	1.66
200	1723.0	0.014	0.02	0.05	0.10	0.21	0.31	0.42

[a] Based on normal life of 180,000h. Time durations not shown are in excess of 24h.
[b] This column of time durations for 0.013% loss of life gives hours of continuous operation above the basis-of-rating hottest-spot temperature (110°C) for one equivalent day of operation at 110°C.

Figure 12.21 - Life of winding insulation versus temperature, 65 °C rise designs. (From IEEE Std. C57.91-1995. Copyright 1995 IEEE. All rights reserved.)

actual transformers. The power transformer curves were based on aging tests of insulation samples in test containers to achieve 50% retention tensile strength. In C57.91-1995, one end of life definition is now 25% retention of tensile strength. Another end of life definition is the decline of the degree of polymerization (DP) number to a value of 200. Tensile strength and DP retention values were determined by sealed-tube aging tests on well-dried insulation samples in oxygen-free oil.

C57.91 allows the user to select the criteria most acceptable to their need. An insulation-aging factor may then be applied. A per-unit life concept and aging factor were introduced in this loading guide. Equations are available to calculate percent loss of insulation life. This then can be applied to estimate end of reliable life. End of reliable life is a point when the transformer has lost its mechanical strength. It is recognized that under certain conditions transformer life can well exceed reliable life.

Effect of Loading Beyond Nameplate Rating

Loading is excess of nameplate rating certainly involves some degree of risk. Aging and mechanical deterioration of winding insulation has been the primary concern in the past. We now know that loading beyond nameplate rating can also lead to:

- Gas bubble evolution from the insulation, produced by high temperature, moisture, eddy-currents...etc.
- Damage to structural insulation
- Deformation of conductor and structure.
- Damage due to pressure buildup.
- Increased resistance in contacts from oil decomposition products in localized high temperature area when the load tap-changer is loaded beyond nameplate rating.
- Problems due to oil expansion.

Transformer Insulation Life

"The life of the transformer is the life of the insulation" and "The life of the insulation is a controllable factor."

Aging or deterioration of the insulation is a function of temperature, moisture, and oxygen levels over time. With modern oil preservation systems available, the oxygen and moisture that contribute to insulation aging can be minimized, if not eliminated, leaving temperature as the primary enemy of insulation life. Since the temperature profile in most transformers is not uniform, most of the aging will occur in the warmest regions of the transformer.

The loading guide C57.91 gives the user many tools to predict transformer insulation aging. It would stand to reason if you have a transformer well-dried (>0.5% moisture content in the insulation) and oxygen free, by monitoring the average daily temperature you could plot the aging rate. For instance, if you have a transformer with an average daily hot spot temperature of 110 °C, the aging rate would be one. If you choose a retained DP number of 200 as your end of life criteria, then it would take 150,000 hours until it reached this level, or 17.12 years.

The Rule of Eight Degrees

As you increase the temperature of the transformer above the designed limits the insulation will age faster. Referring to the loading guides, if we increase from a daily average temperature from 110 °C to 118 °C, the aging factor will increase to 2.23 (Table 12.3)! Conversely if we operate the transformer at an average daily temperature of 102 °C (8 °C cooler) the aging factor would be 0.434! This would increase the insulation life of a well-dried, oxygen-free system to approximately 39 years! Or instead of using 24 hours of its life up per day at 110 °C, it would use 10.4 hours of its life in a 24 hour period.

Determining transformer reliable life is not an easy task. The loading guide will provide knowledge and guidance in your decision-making process with regards to load cycles, short-time emergency loading, long-time emergency loading, operating with diminished cooling capacity, effects on other components, unusual temperature and altitude conditions, cold load pick-up and much more.

NOMEX® Insulated Conductor

For a reasonable increase above the cost of conventional cellulose insulated conductor, many benefits associated with the properties of NOMEX® conductor insulation can be realized. Benefits include increased thermal, electrical, and mechanical capabilities for your power transformer rewinds.

Aging Acceleration Factor

60°C	.0028	80°C	.0358	100°C	.3499
62°C	.0037	82°C	.0455	102°C	.4337
64°C	.0048	84°C	.0577	104°C	.5362
66°C	.0062	86°C	.0729	106°C	.6614
68°C	.0080	88°C	.0919	108°C	.8142
70°C	.0104	90°C	.1156	110°C	1.000
72°C	.0134	92°C	.1149	112°C	1.2256
74°C	.0172	94°C	.1813	114°C	1.4990
76°C	.0220	96°C	.2263	116°C	1.8296
78°C	.0281	98°C	.2817	118°C	2.2285

Table 12.3

Thermal Considerations

The high temperature stability of NOMEX® (hot spot temperatures of 200 °C can be tolerated) results in negligible insulation aging at normal or overload conditions. NOMEX® insulation will not embrittle, and therefore, decomposition products will not affect the life of the total dielectric system. NOMEX® insulation will not degrade transformer oil with the release of CO, CO_2 and H_2O. NOMEX® conductor insulation is insensitive to moisture, having less than half the moisture absorption of cellulose.

Electrical Considerations

The improved electrical properties of NOMEX® over cellulose provide an extra margin of safety. Published literature suggests ratio values of (1500 vs. 1200) for 60 Hz Volts/mil and (3000 vs. 1800) for impulse Volts/mil. The lower dielectric constant of NOMEX® results in better stress distribution between solid and liquid insulation.

Mechanical Considerations

NOMEX® has superior physical properties. Its abrasive and tearing strength is quite high in comparison to cellulose-insulated conductor. The resilience of NOMEX® resists shrinkage, therefore, tighter coils result that provide continued mechanical integrity under short circuit force even after years of service.

Many transformer applications and duty cycles require special considerations. Increased electrical and mechanical margins, along with virtually eliminating all anxiety for conductor insulation aging, can raise your comfort level when making an important decision.

Chapter 13 | Cooling Systems

Purpose and Function

A transformer's core and coils generate heat through I^2R losses. Heat shortens a transformer's life. One of transformer oil's main functions is to transfer heat away from the core and coils. The transformer's cooling system's main function is to transfer this heat from the oil into the outside air, thus cooling the transformer, extending its reliable life, and increasing the load it can carry.

Types and History

"Yukon" coolers increase the tank's surface area, offering moderately more cooling than just the flat tank walls do. Yukon coolers still appear on modern transformers that require only a little more oil cooling than the transformer tank alone provides. (Figure 14.1)

Figure 14.1 - Yukon Cooler

Figure 14.2 - Tube style

Tube radiators further increase surface area. (Figure 14.2) They also provide opportunity for a convection cooling cycle that increases heat transfer away from the core and coils through the oil to the outside air. As the hot oil rises inside the transformer, the cooled oil sinks inside the radiators. (Figure 14.3).

Flat-tube radiators (Figure 14.4) provide more surface area than round-tube radiators (and so exchange more heat).

Flat-fin (pancake) radiators (Figure 14.5) provide more surface area than flat-tube radiators.

The next step to increase cooling capacity is the addition of fans to increase the flow of air across the radiators' cooling surfaces.

The step to increase cooling capacity beyond that provided by fans is the addition of oil pumps to increase the flow of oil through the transformer's core and coils and through the radiators or heat exchangers.

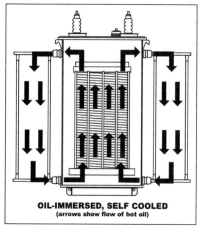

Figure 14.3 - Oil-immersed, self cool system - arrows show flow of hot oil.

Figure 14.4 - Flat-tube style

Figure 14.5 - Flat-fin pancake style

Nameplate Designations

Over the years, many different nameplate designations for cooling systems have been used. Some of the more common include:

- OA – Oil Air cooling. Includes all liquid-filled transformers with or without radiators that don't have fans or pumps.

- OAFA – Oil, Air, Forced Air cooling. Includes all liquid-filled transformers with radiators and fans.

- OA/FFA – Oil Air, Future Forced Air. Includes all transformers with radiators, no fans, but with the internal capacity for increased load with a greater cooling capacity. These transformers can be up graded with fan packages or other additional oil cooling and run at a higher capacity.

IEEE suggests a new, four-letter classification system for 2000 and beyond.

- The first letter designates the internal cooling medium:

    O for oil or a synthetic fluid with a fire point < 300 °C
    K for above but with a fire point > 300 °C
    L for insulating fluids with no measurable fire point.

- The second letter designates the how the insulating liquid is circulated through the cooling system AND through the transformer:

    N for natural convection through the radiators and tank
    F for forced circulation through the cooling system
    D for forced circulation through the cooling system AND directed flow from the cooling system into the tank and windings.

- The third letter designates the cooling medium outside the transformer:

    A for air
    W for water

- The fourth letter designates how the cooling medium outside the transformer circulates:

    N for natural convection
    F for forced air (fans) or forced water (pumps).

Adding Additional Cooling Capacity

Due to High Ambient

If a transformer operated at nameplate capacity overheats due to high ambient temperatures (warm climates) or poor air circulation (vaults), the transformer owner may choose additional oil cooling capacity. This could take the form of fans (Figure 14.6) or an external oil cooling unit (Figure 14.7).

To Increase Transformer Capacity

Unless a transformer was designed to accept additional cooling capacity beyond what the manufacturer shipped it with, it probably should not be given additional cooling in order to increase its load capacity.

For example, a transformer designed to operate at 1000 kVA will generate a certain amount of heat. The transformer manufacturer designed oil flow, spacing, and cooling ducts to transfer away from the core and coils the amount of heat generated by a 1000 kVA load. The transformer's leads and tap changer and were designed to carry 1000 kVA of current. The transformer's bushings

Figure 14.6 - External cooling system (courtesy of Unifin International)

Figure 14.7 - Cool-a-tran - an external oil cooling unit.

and internal insulation were also designed to carry 1000 kVA of current. Simply adding extra oil cooling capacity to operate the unit at, say, 1500 kVA will not necessarily improve oil flow through the core and coils where it is most needed. Extra cooling may suppress the top oil temperature while allowing the hot spot temperature to climb beyond the safety zone. Overall coil insulation temperature could also increase, using up insulation life at a rate the transformer owner would find unacceptable. The additional current would also shorten the reliable operating life of other conductors such as the Tap Changer and the bushing leads.

Transformers with FFA ratings were designed and manufactured taking into account all these considerations for carrying increased current. Adding increased oil cooling capacity up to the higher designated load level will not stress these transformers beyond safe operating parameters.

Chapter 14

Tank and Paint

Purpose and Function

Protective coatings or paints are used for various purposes. In most homes and office interiors, for example, they are primarily used for aesthetics, whereas, in steel structures such as chemical plants, coastal bridges, electric transmission towers, and apparatus surfaces, they are used for corrosion resistance to the atmospheric environment. Without coatings, such steel surfaces would corrode rapidly and revert to iron oxide.

Coatings provide protection against deterioration by forming a very thin protective layer (or film, 6-11 mills) that must resist all destructive atmospheric elements. Proper selection, maintenance and repair of this thin film can extend the life of protected equipment. Such maintenance and repair of coated surfaces can also result in enormous financial savings, if performed at the appropriate time. Thus, a minor touch up of the protective coating may save apparatus from disintegrating beyond repair to a state where the whole structure will require a complete new coating system - a task involving thorough surface preparation, priming, and finish coating. Therefore, in selecting and applying protective coatings, it is important to consider the quality of coatings to be employed.

Historically, any quality coating available was applied to electrical apparatus substrates. This practice is inadequate for today's computer-designed and hotter running transformers. High-performance coatings must serve a dual purpose. First, the paint system must allow for rapid radiation of heat, and be capable of reflecting solar light rays. Second, the system must protect against localized atmospheric conditions.

Since transformers generate internal heat, they must be properly coated to help prolong their lives. Color affects a transformer's ability to do this. Therefore, transformers require a different evaluation of paint systems because of the multiplicity of problems to be solved. Among the requirements for pad-mounted transformers (ANSI Standard C57.12.22) is an undercoating over the regular finish shall be applied to all surfaces that are in contact with the pad to minimize corrosion.

At least 20 different systems of painting transformers are now on the market. Considering the variables of each of these systems, an infinite number of combinations are available from which a choice can be made. Manufacturers often select a system and a color that are adequate for most applications without being specific to many. Some transformer buyers immediately repaint their transformers. A layman can scarcely take time to research all of these new products as they come on the market. It is important, therefore, to seek professional help from coating control engineers who can help to control corrosion in transformers.

The Corrosion Problem

Of all the metals available, the principal metallic substrate of electrical apparatus is steel, because of economy, flexibility, and ease of fabrication; nevertheless, it is the most likely metal to corrode.

The most minute form of coating deterioration often leads to pinhole rust. This is followed by underfilm cutting, which destroys the coatings surrounding the rust blooms. These rust blooms can quickly develop into isolated pitted rust if there is a direct chemical attack on steel surfaces. With time, rust penetration is accelerated, leading to oil-leaking problems, particularly in the area of the cooling fins, tubes, and weldment areas.

When such a condition occurs, the overall coating may not be salvageable. The delay which occurred in preventive maintenance will be very costly if repair to cooling radiators and downtime is needed. Therefore, the best method for corrosion control is to attack the problem when the initial signals of deterioration occur.

Since the cost of corrosion and corrosion control to the United States has been estimated at nearly 70 billion annually, corrosion control of apparatus should be an indispensable goal of engineering.

Corrosion is normally caused by chemical and electrochemical oxidation, and occurs in the presence of both oxygen and water.

Dampness - The Overriding Factor in Corrosion

Most contaminants will have little or no effect in the absence of moisture. A film of dew, saturated with corrodents, provides a very aggressive electrolyte for the promotion of corrosion.

Humidity above 80% causes rapid rise of corrosion. This is due to the nature of the hygroscopicity of contaminants. A constant source of wetting is provided to cracks, crevices, cooling fin bracing, and under top-lids of transformers. It has been said that two-thirds of all continental corrosion occurs in humid climates near oceanic bodies.

Thus, a damp surface attracts corrosion products from the air, and provides a basis for accelerated corrosion. Among the major types of corrosion are the following: bimetallic, crevice, pitting, thermogalvanic, spalling, and poultice (internal, acidic oil, sludges) corrosion. Such forms of corrosion are often associated with rural, marine, caustic, and chlorinated environments. One of the worst environments relative to dampness is a location downwind from a cooling tower.

Since corrosion is impractical to eliminate, the secret to sound engineering is its early detection and control through barrier coatings, rather than preventing it.

Selecting the Right Coating System

Two major factors come into play in deciding what kind of coating system should be chosen - environment and color.

Choosing the right coating system for electrical equipment is not an easy task. It may be better described as a process of elimination rather than selection. Today, we are swamped with paint choices - epoxies, vinyls, oil coatings, alkyd resin paints, phenolic resin paints, urethanes, etc. These are only a few of the many different types of coatings available for maintenance painting. Remember, there are over 20 different systems with variables in each system. Therefore, in choosing a coating system, particularly for transformer units, one should be guided by certain criteria. Table 15.1 illustrates criteria that can be employed in selecting coatings for transformers.

Environmental Criteria

Two additional factors should also be taken into consideration - the amount of surface preparation required, and the environment in which the equipment is located. Selecting protective coatings for different environments may be determined by the geographical location (see Table 15.2). This is important because there is a direct relationship between color and operating temperature. In other words, color can have some effect on transformers in service.

Effect of Color on Paint Systems

In choosing a surface color, one should consider the effect of internal heat loads on the paint system. This is important since certain colors will increase temperature rise of transformers in sunlight.

Traditionally, electrical equipment has been painted dark colors such as black, blue, green, and gray. The 50-year-old

Criteria for Choosing Transformer Coatings*

Barrier Coating Type (finish coats)	Gloss retention	Color retention	Fading	Water resistant	Acid resistant	Alkali resistant	Solvent resistant	Marine resistant	Adhesion	Hardness	Temperature 190°C
Alkyd	A	A	A	A	NR	NR	NR	NR	V	A	V
Alkyd silicone	V	V	V	V	NR	NR	A	V	V	A	V
Vinyl	A	A	A	V	V	V	AV	V	A	VA	NR
Polyester epoxy	A	V	A	V	A	V	V	V	V	V	V
Polyamide epoxy	N	A	A	V	A	V	V	V	V	A	V
Amine epoxy	NR	NR	NR	V	V	V	V	V	V	V	NR
Oil-modified epoxy	NR	NR	NR	V	A	A	A	V	A	V	A
Chlorinated rubber	A	A	A	V	V	NR	A	A	A	A	NR
Urethane (ASTM Class V)	V	V	V	V	V	V	V	V	A	V	A
Oil-modified urethanes	A	A	A	A	A	NR	A	A	A	A	NR

*Relative resistance properties of commonly used transformer coatings.

V = Very Good, A = Average, NR = Not Recommended

Primer Coating Type	Adhesion	Chemical resistant	Hardness
Long oil	V	NR	NR
Vinyl Alkyd	V	A	A
Epoxy	V	V	V
Water	A	A	A

Table 15.1

Selecting Protective Coatings for Different Environments

Industrial	Pollution/Smog	Polyester epoxy coating Vinyl/alkyd primer
Disastrous	Acids/Bases combined with humidity, etc...	ASTM class V polyurethane Epoxy Primer Acid Pre-primer.
Rural	Sun, Wind, Rain	Alkyd silicone enamel Long-oil primer
Marine	Humidity, Fog and Salt	Alkyd silicone enamel

Table 15.2 - Environmental Criteria

theory held that while white bodies absorbed less solar heat than black bodies, black bodies tended to radiate internal heat better than white bodies. The practical assumption was that the two effects neutralize each other. An operating temperature rise of a white-painted transformer was only about 10 to 20% lower than a black-painted unit. So, the actual difference of no more than 2 °C advantage could not justify the use of special paint, meaning other than the standard black. However, today's operating temperatures have increased by a factor of two or more, making that 10 to 20% reduction more attractive.

White bodies have proven to radiate low-temperature (below 200 °C) heat equally as well as black bodies and they reflect solar radiation. Up to 200 °C, both dark and white colors absorb 95% heat. Hence, in the range transformers normally operate, white will conduct with the same rapidity as a dark color.

Figure 15.1 shows that white bodies do not absorb heat well at high temperatures. Thus, contrary to past theory (black body emission), a white body will operate several degrees cooler than a dark-colored surface in the temperature range of operating transformers. Reflectance is the ability of the top coat to reflect sunlight, and consequently heat. Gloss reflectivity can only achieve its maximum potential by using hard and smooth-surfaced finish coating.

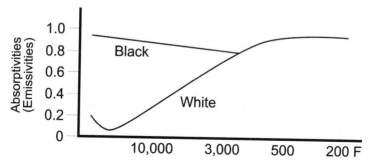

Figure 15.1 - Comparing the emissivity of black and white paints.

Silicone alkyds, polyester epoxy, and ASTM class V polyurethane exhibit the highest order of gloss and color retention. Table 15.3 shows how the various colors compare in terms of light reflection.

To illustrate the effect of color on paint systems, an example will suffice. In climates with direct sun radiation, a black-painted transformer under extremes of solar radiation would increase the temperature 15 °C above the ambient temperature generated by the equipment. On the other hand, a white-painted unit will absorb a little radiation (about 15% of 15 °C or 2 °C). Therefore, a white transformer operates 13 °C cooler. Table 15.4 compares various colors and their radiation absorption coefficient.

In choosing a "light colored" paint, one must not make the mistake of picking aluminum paint. Studies have shown that although aluminum has good sun reflectance, (up to 40 °C) the aluminum flakes from such paint overlap like leaves, and form an insulating barrier which slows dissipation of internal heat (aluminum emissivity is 0.55 at normal temperatures). A plain aluminum-painted tank will run approximately 30% higher temperature rise than if painted with a non-metallic paint.

Color	% Color Reflection
White	80-97%
Very Light Tints	70-80%
Light Tints	60-70%
Medium Tints	25-60%
Aluminum	41%
Dark Blue	3-15%
Black	1-2%

Table 15.3

Paint Color	Solar Radiation Absorption Coefficient
White	0.14
Cream	0.25[1]
Yellow	0.30[1]
Light gray	0.35[3]
Light gray, green, blue	0.50[1]
Medium gray, green, blue	0.75[1]
Dark gray, green, blue	0.95[1]
Black	0.97[2]

[1]Approximate - because of wide variations of the color spectrum.
[2]Variance in black colors.
[3]Estimate.

Table 15.4

Exploring High-Performance Coatings

Since no paint is suitable for all applications, it is necessary to tailor the product to specific degrees of corrosion.

Mild, Marine or Rural

The use of alkyd-silicone enamels is very evident where lasting appearance, resistance to mild chemical exposure and marine atmosphere prevail. The initial gloss approaching 90% provides for the reflectance of solar light rays. The presence of 30% silicone is the preferred formula.

Heavy or Chemical

Catalyzed polyester epoxies are formulated for optimum gloss and color retention qualities in their generic family. The hard and glossy finish results in a coating which is resistant to chemical attack, abrasion and high heat. While some epoxies will quickly fade and discolor, the polyester formula will retain its appearance longer in constant temperature ranges exceeding 200 °F (93 °C).

Corrosive Chemical Atmosphere

ASTM Class V polyurethanes, when coupled with a catalyzed epoxy primer, are gaining wide usage wherever corrosion engineers specify the most physically durable, flexible coating. They are used by airlines to protect their jets, by railroads for their diesel-electrics, and the transportation industry selects this coating in corrosive areas where maximum resistance to chemical attack is mandatory. No coating is without several negatives. ASTM Class V polyurethane is no exception. Negatives include:

- Oxygenated solvents (including alcohols).

- TDI or HDI (solvents in formula possibly may cause bronchial discomfort and if inhaled for extended periods may cause bronchialnoma or cancer of the lungs).

- Difficulty in flow-coating.

- When being sprayed, eye irritation and tears will result.

Ongoing Research

Environmental demands and governmental concern for human health have intensified research and development efforts of the coating industry.

One focus is on the refinement of water-soluble coatings. At this point, water-borne coatings do not have the proven durability of solvent-based coatings. During the interim, however, proper corrosion control starts with preventive maintenance - the recoating of units before rust becomes too apparent. Even new transformers should be carefully examined after installation. Scratches and scuffs frequently received should be touched-up with paint.

Chapter 15

Bushings

Purpose

Conductors to and from the transformer windings must safely be brought through the tank walls without letting the current go to ground. The bushing insulator performs this function. Transformer bushings are designed to insulate a conductor from the transformer tank. In addition to being an effective insulator, bushings must also be water-, gas-, and oil-tight to keep moisture out of the transformer.

Design

In order to increase the "creepage" distance between the conductor and the tank, the distance over the outer surface is increased by adding "petticoats" or "watersheds." The watersheds also reduce potential flashover across the surface due to moisture or particulate contamination.

A clean, dry dielectric material inside the bushing surrounds the conductor being insulated. Transformers rated 15 kV and higher, typically would have condenser bushings as standard equipment (Figure 16.1). Oil-impregnated paper, alternated with metallic foil or conductive ink paper, wraps around the conductor core in condenser bushings. The foil layers or conductive ink wraps form a series of condensers that equalize voltage distribution between the outer and inner layers of insulation.

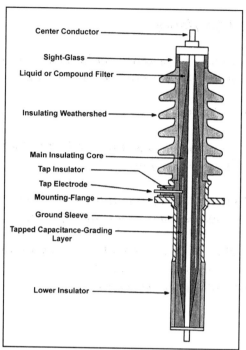

Figure 16.1 - Oil-Filled and Condenser-type bushings (courtesy of Doble Engineering).

Materials

Transformer bushings traditionally have been externally clad in porcelain due to porcelain's excellent electrical and mechanical qualities. More recently molded, solid silicones have been used successfully for transformer bushing housings.

Internally, materials such as porcelain, toughened glass, fiberglass, epoxy or some other cast organic material, resin-bonded paper or resin-impregnated paper enclose the conductor. Porcelain insulators are generally oil-filled beyond 35kV.

The voltage level to be insulated determines what bushing type is required.

What Could Go Wrong

Transformer failures caused by bushing failure come in second only to transformer failures originating in the windings.

Bushings fail when:

- Current finds its way to ground across the external surface of the bushing.

- Current finds its way from inside the bushing due to condenser failure through to ground.

The first failure mode occurs due to non-dielectric contamination on the external surface of the bushing. The second mode occurs due to some type of material breakdown.

Appendix A
Acronym Identification

ANSI
American National Standards Institute
1430 Broadway, New York, NY 10018

ASTM
American Society for Testing and Materials
100 Barr Harbor Dr., W. Conshohocken, PA 19428

IEEE
Institute of Electrical and Electronics Engineering
445 Hoes Lane, Piscataway, NJ 08854

NEMA
National Electrical Manufacturer's Association
2101 L Street, Washington, DC 20037

IEC
International Electrotechnical Commission
1, Rue de Varembe, Geneve, Switzerland

EPRI
Electrical Power Research Institute
P.O. Box 10412, Palo Alto, CA 64303

EEI
Edison Electric Institute
111 19th Street SW, Washington, DC 20036

NETA
International Electrical Testing Association
P.O. Box 687, Morrison, CO 80465

Index

[A]
absorbent, 68
acceptance tests, 163, 166, 183
acid number, 88, 288
adsorbent, 68, 298
alpha, 11
aluminum, 4, 320
arcing, 132, 146, 149
Arrhenius equation, 61
Askarel testing, 75

[B]
basic insulation level (BIL), 24
Buchholz relay, 116, 127
bushing power factor, 187, 195
bushings, 237, 429

[C]
capacitance, 11, 193
catalyst, 62, 64, 68, 69
cellulose, 68, 103, 133, 265, 271
coatings, 221, 419, 423
coil blocking, 16
color test, 96
conservator, 20
cooling, 18, 411, 414
copper, 4, 320
core, 321, 349
core ground, 187, 213, 356
corona (partial discharge) 132, 148
corrosive sulfur, 159, 162
critical transformer, 74, 292

[D]
DBPC, DBP, 86, 289
degasification, 280
degree of polymerization, 107, 377, 382, 404
dehydration, 271
dew point, 46
dicyandiamide, 21
dielectric (defined), 51
dielectric absorption, 187, 205
dielectric breakdown voltage (D 1816), 99
dielectric breakdown voltage (D 877), 97
dielectric constant, 81, 383
dielectric strength, 55
disc winding, 5, 9
disposal, 327, 332
dissipation factor, 78
dissolved gas analysis (DGA), 24, 38, 49, 101, 113, 130, 171, 178
dissolved metals analysis, 102
Dornenberg ratio method, 140
Duval triangle, 141
dynamic restraint, 17

[E]
eddy current, 344, 357
electrical testing, 36, 177, 181
esters (natural and synthetic) testing, 76
excitation test, 187, 197

[F]
factory electrical tests, 36, 181
fault gas detection, 117, 129
faults, 132, 137
field assembly, 44
filter press, 297
Fluidex, 305
free breather, 20
frequency response analysis, 217
fuller's earth, 299, 305
functions of oil, 19, 53
furaldehyde, 104
furans, 103, 105
furans and DP, 109

[G]
gas blanket, 20, 130
gas chromatography (GC), 109, 113, 117, 123
gassing tendency, 159, 161
ground resistance, 215

[H]
heat transfer, 55
high performance liquid chromatography (HPLC), 117
Hot Oil Clean®, 263, 293, 307
hot spot overheating, 132, 145, 150
hysteresis, 358, 359

[I]
ICP, 102
infrared (thermography), 178, 247
infrared spectrophotometer, 86
inhibitor, 86, 162, 307
in-service oil standards, 32, 284
installation, 40
insulation resistance, 187, 202, 388
insulation system, 14, 398, 403
InsulDryer, 280
interfacial tension, 91, 288
interpreting DGA data, 136, 154

[K]
Karl Fischer moisture analysis, 81, 171, 187
key gases, 138, 154
Kraft paper, 15, 21, 54, 378

[L]
layer winding, 5, 7
leak test, 46, 265
leakage current, 203
leakage reactance, 216

lightning, 25, 367
lightning impulse, 369, 372
liquid power factor, 78, 165, 187, 197
load loss, 13, 360
load tap changer (LTC), 24, 79, 111, 167

[M]
maintenance recommendations, 112, 263, 289
major insulation, 15
metals in oil, 102
mineral oil (defined), 52
mineral oil specification, 31, 157, 159
minimum insulation resistance, 204, 269
minor insulation, 15
moisture by dry weight, 3, 374
moisture content, 81, 380

[N]
neutralization number, 88
nitrogen system, 20
no load loss, 357, 360
Nomex®, 408

[O]
oil preservation, 19
oil screen tests, 74, 87, 170
overpotential testing, 187, 212
oil testing, 71
oxidation, 57, 283, 383
oxidation stability, 62, 160, 162
oxidation by-products, 57, 58, 69
oxygen, 60, 399

[P]
paint, 221, 419
particle count, 110, 168

particles and filming compounds analysis, 111, 168
PCB, 109, 327, 332
percent moisture by dry weight, 3, 82, 84, 271, 379
percent saturation, 82, 83, 390
perchloroethylene testing, 76
Piper chart, 279
polar compounds, 58, 301
polarization index, 187, 205, 207
powder blasting, 225, 253
power factor, 184, 187, 188, 191
power factor tip up test, 187, 194, 269
power factor valued oxidation, 162
primers, 228
pyrolysis, 133, 134

[R]
radial force, 371
rates of gas evolution, 135
reaction rate, 61, 70
reclaiming, 283, 294, 298, 303
reconditioning, 293, 296
recovery voltage method, 219
regeneration, 283, 294
reinhibiting, 286, 288, 294, 305
relative density (specific gravity), 95
resistivity, 81
retrofilling, 298, 309
rewind, 319
Rogers ratio method, 140, 156
rule of eight degrees, 19, 401, 407

[S]
sampling, 118
saturation, 82, 83
shipping, 41
silicone testing, 76

sludge, 56, 283, 293, 295, 400
sludge free life, 62
solid insulation, 14
solubility, 83, 390, 393
sparking, 132
specific gravity, 95
standards, 26, 30, 172, 185, 223, 365
static restraint, 16
step voltage, 187, 207
stray gassing, 161
switching surge, 369, 373
synergy, 65
syringe, 119

[T]
tan delta (δ), 67
ten degree Celsius rule, 61, 62, 40
tensiometer, 91
test methods, 35
testing packages, 74
turns ratio test, 187, 208

[U]
upgraded insulation, 21

[V]
visual test, 97
vacuum extraction, 121, 122
vacuum filling, 46, 277
vertical force, 370

[W]
warranty, 49, 163
winding insulation, 15
winding resistance, 187, 210
windings, 4, 5, 7, 335